General Physics and Heat

George A. Grant

Lecturer in Physics, Hatfield Polytechnic

Edward Arnold

© G. A. Grant 1977

First published 1977 by Edward Arnold (Publishers) Limited
25 Hill Street, London W1X 8LL

Paper edition ISBN 0 7131 2623 X

To Shirley

Photoset and printed in Malta
by Interprint (Malta) Ltd

Preface

The typical physics text book deals with many topics in more detail than is necessary for the average first-year college or university student. Such text books may be ideal for the student who is approaching his finals and already has a good grasp of the fundamentals of physics, but for the student who has just completed 'A' levels and may not be intending to specialise in physics, a more gentle introduction is required to many topics, some of which may be quite difficult.

This book is written to provide such an introduction. An average student with a knowledge of 'A' level physics and mathematics should have no difficulty in following the work, even without assistance from lectures and tutorials. Having mastered the contents of this book, the student will be able to proceed with confidence to more advanced topics.

Sixth formers who wish to further their knowledge of physics before going on to university, or who are preparing for scholarship papers, first- and second-year physics degree students and students studying physics in teacher training colleges should all find this book useful. Other students for whom some physics is necessary – engineering students, medical students etc. – will also find this book helpful.

Only general physics and heat are covered since it would not be possible to cover all physics topics thoroughly in one volume. The aim of this book is to explain the fundamentals of the subject and to provide sufficient information to stimulate the student to read elsewhere for specific methods and experiments.

I am grateful to Dr A. R. Stokes of King's College, University of London, for helpful advice and comments during the preparation of this book and also to Dr J. Houghton of the Hatfield Polytechnic. Finally I am indebted to the late Dr Humphries-Owen, and to the late Professor W. Ehrenberg, both formerly of Birkbeck College, University of London, whose lectures, which I attended as a student, introduced me to the physics covered in this book.

G. A. Grant

Hatfield,
1977

Contents

Part 1   General Physics

1 Dynamics of a Rigid Body

Before considering a rigid body we shall review briefly circular motion and simple harmonic motion of a particle.

1.1 Uniform motion in a circle

Angular velocity $\omega = d\theta/dt$ rad s^{-1}, where t denotes time. Period $T = 2\pi/\omega$, frequency $f = 1/T = \omega/2\pi$. The tangential velocity v is related to the angular velocity ω by $v = r\omega$ where r is the radius of the circle.

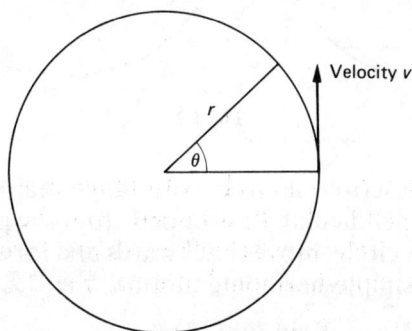

Fig. 1.1

When a body moves in a circle with uniform velocity v it experiences an acceleration towards the centre of the circle of magnitude v^2/r or $r\omega^2$.

1.2 Simple harmonic motion

This is the motion of a particle whose acceleration is proportional to its distance x from a fixed point and is always directed towards that point.

Acceleration (in direction of x increasing) $= -\omega^2 x$

where ω is a constant. The general solution of this equation is

$$x = a \begin{cases} \sin(\omega t + \delta) \\ \cos(\omega t + \delta) \end{cases}$$

where a and δ are arbitrary constants and t denotes time. For a particle moving with angular velocity ω

$$x = a \cos(\theta + \delta) = a \cos(\omega t + \delta)$$

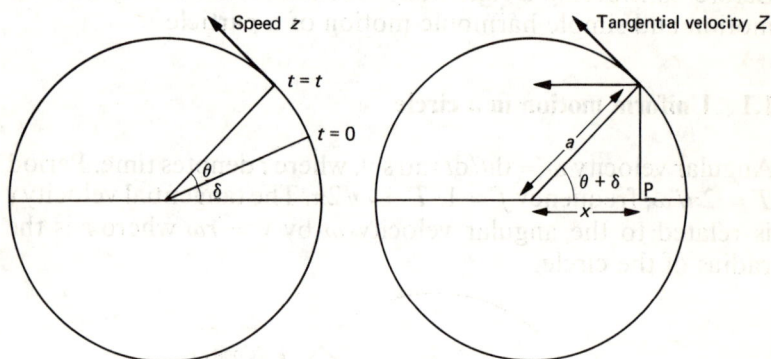

Fig. 1.2

As a particle describes a circle with tangential velocity Z, the foot of the perpendicular P, dropped from the particle on to a diameter of the circle, moves backwards and forwards along the diameter with simple harmonic motion, Fig. 1.2.

Velocity of P $= -Z \sin(\theta + \delta)$
$\qquad\qquad = -a\omega \sin(\theta + \delta)$

Velocity $= \dfrac{a\omega\sqrt{a^2 - x^2}}{a} = \omega\sqrt{a^2 - x^2}$

Period of oscillation of P, $T = 2\pi/\omega$

Acceleration $= -\omega^2 x$

Displacement, $x = a \cos(\omega t + \delta)$

Velocity at any instant $v = \omega\sqrt{a^2 - x^2}$

Maximum velocity $= \omega a$

The maximum displacement a is the amplitude of the motion,

1.3 ROTATION OF RIGID SOLID BODIES

δ is the phase constant, $\omega t + \delta$ the phase angle.

Maximum kinetic energy $= \frac{1}{2}m\omega^2 a^2$

where m is the mass of the particle.

Maximum potential energy $= \int_0^a m\omega^2 x \mathrm{d}x = \frac{1}{2}m\omega^2 a^2$

1.3 Rotation of rigid solid bodies

A rigid body is a body in which the distance between any two component parts remains constant. Consider a small mass m,

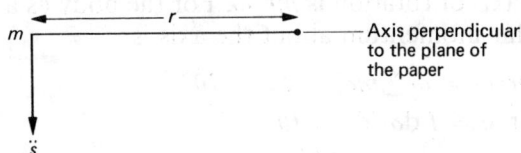

Fig. 1.3

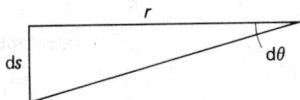

Fig. 1.4

part of a rigid body, which is given an acceleration $\mathrm{d}^2 s/\mathrm{d}t^2 = \ddot{s}$ normal to the radius r by some force. The partial force, f, is given by

$$f = m\ddot{s}$$

But $\mathrm{d}s = r\mathrm{d}\theta$

thus $\mathrm{d}s/\mathrm{d}t = \dot{s} = r\dot{\theta}$

and $\mathrm{d}^2 s/\mathrm{d}t^2 = \ddot{s} = r\ddot{\theta}$

therefore $f = mr\ddot{\theta}$

The partial couple, also known as the torque or moment, is given by

$$f \times r = mr^2\ddot{\theta}$$

For the body as a whole

$$\text{total couple} = \ddot{\theta}\sum mr^2$$

$\sum mr^2$ is called the moment of inertia, I, about the particular axis being considered.

$$\text{Couple} = I\ddot{\theta}$$

If I is large then a large couple is required to produce a given acceleration. The tangential velocity of the mass m perpendicular to r is $r\dot{\theta} = r\omega$, thus

$$\text{linear momentum} = mr\omega$$

The moment of this momentum, or the angular momentum, about the axis of rotation is $mr^2\omega$. For the body as a whole the total angular momentum about the axis is

$$\sum mr^2\omega = \omega \sum mr^2 = I\omega = I\dot{\theta}$$
$$\text{couple} = I\, d\omega/dt = I\ddot{\theta}$$

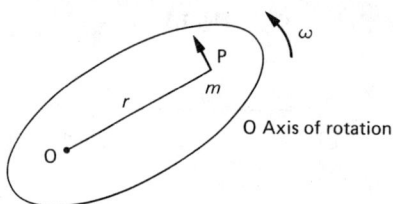

Fig. 1.5

The velocity of P is $r\omega$ and the kinetic energy of the particle of mass m is $\frac{1}{2}mr^2\omega^2$.

The kinetic energy of the whole body is

$$\tfrac{1}{2}\omega^2 \sum mr^2 = \tfrac{1}{2}I\omega^2$$

1.4   Calculation of moments of inertia

Cylinder
The moment of inertia of a cylinder about the axis of symmetry (see Fig. 1.6) is given by

$$I = \int_0^R \rho.2\pi r dr.Lr^2$$

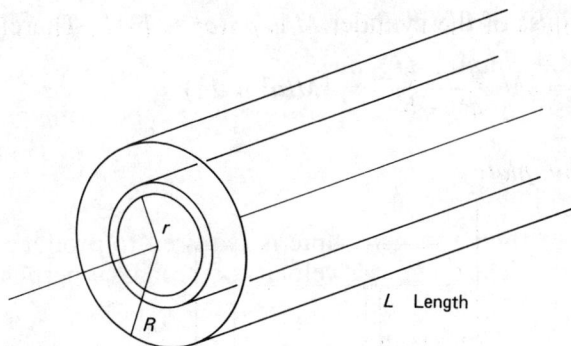

Fig. 1.6

where ρ is the density. Thus

$$I = \int_0^R 2\pi\rho Lr^3 \,\mathrm{d}r = \tfrac{1}{2}\pi\rho LR^4$$

But the mass of the cylinder M is $\rho\pi R^2 L$, therefore

$$I = \tfrac{1}{2}MR^2$$

$\frac{1}{2}MR^2$ is also the moment of inertia of a thin disc about an axis through the centre of the disc perpendicular to its plane.

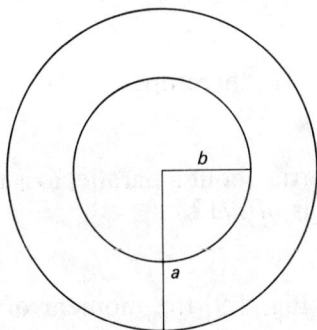

Fig. 1.7

Consider now a hollow cylinder; from Fig. 1.7 we have

$$I = \int_b^a 2\pi\rho Lr^3 \,\mathrm{d}r$$
$$= \tfrac{1}{2}\pi\rho L(a^4 - b^4)$$

But the mass of the cylinder M is $\rho\pi(a^2 - b^2)L$. Therefore

$$I = \tfrac{1}{2}M \; \frac{a^4 - b^4}{a^2 - b^2} \; = \tfrac{1}{2}M(a^2 + b^2)$$

Rod or thin plate

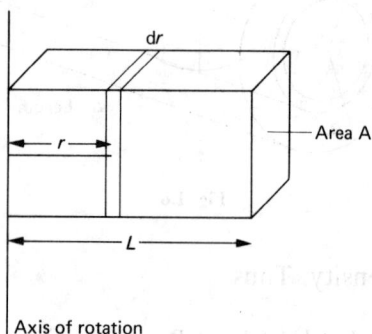

Fig. 1.8

From Fig. 1.8 we see that

$$I = \int_0^L \rho A \,dr.r^2$$

where ρ denotes density. Thus

$$I = \rho A L^3/3$$

But the mass M is ρAL, therefore

$$I = \tfrac{1}{3}ML^2$$

The moment of inertia about a parallel axis through the centre of the rod or plate is $ML^2/12$.

Sphere
With reference to Fig. 1.9, the moment of inertia about the centre of the sphere is given by

$$I_0 = \int_0^R \rho.4\pi r^2 \,dr.r^2 = 4\pi\rho R^5/5$$

But the mass of the sphere M is $4\pi R^3\rho/3$, hence

$$I_0 = 3MR^2/5$$

It should be noted that this is not a moment of inertia in the

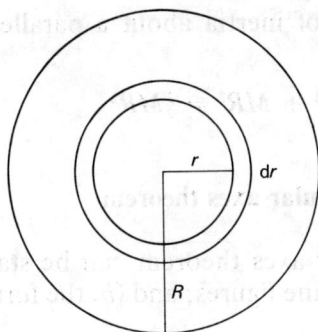

Fig. 1.9

usual sense of the expression. For more difficult axes (involving difficult integrals) two theorems are used (see §§ 1.5 and 1.6).

1.5   The parallel axes theorem

If the moment of inertia of a body of mass M, about an axis through its centre of mass, is I_g, the moment of inertia about a parallel axis at a perpendicular distance a from the first axis is $I_g + Ma^2$.

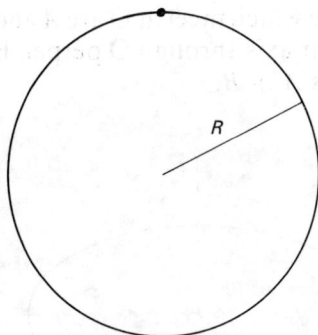

Fig. 1.10

The moment of inertia of a thin disc about an axis through its centre perpendicular to the plane of the disc is $\frac{1}{2}MR^2$. Therefore the moment of inertia about a parallel axis through the circumference is

$$I_{circ} = \tfrac{1}{2}MR^2 + MR^2 = \tfrac{3}{2}MR^2$$

1.6 The perpendicular axes theorem

The perpendicular axes theorem can be stated in two forms: (a) the form for plane figures; and (b) the form for solid figures.

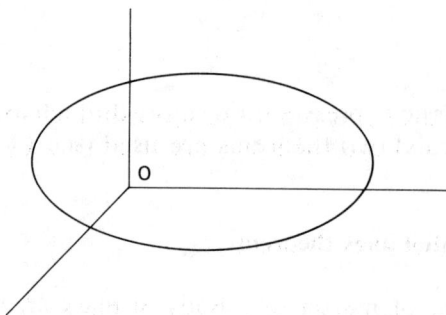

Fig. 1.11

(a) If the moments of inertia of a lamina about two perpendicular axes in its plane which meet at O are A and B, the moment of inertia about an axis through O perpendicular to the plane of the lamina is $A + B$.

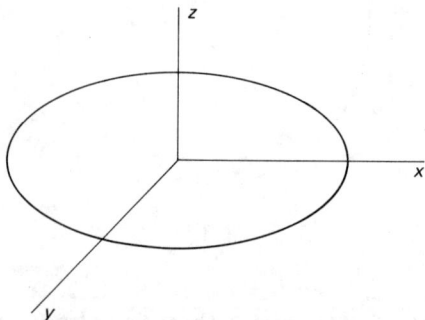

Fig. 1.12

In Fig. 1.12, if I_x, I_y and I_z are the moments of inertia in the x, y and z directions respectively, then from the perpendicular axes theorem we have

$$I_z = I_x + I_y$$

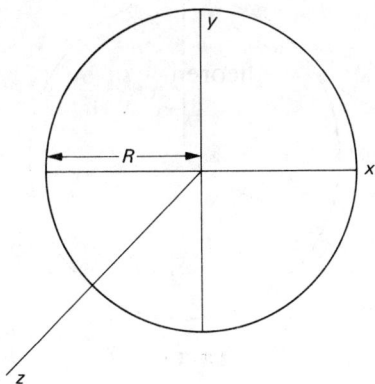

Fig. 1.13

From Fig. 1.13, considering a thin disc, we have

$$I_y = I_{\text{one diam}} \quad I_x = I_{\text{another diam}}$$

and $\quad I_y + I_x = I_z$

But $\quad I_y = I_x$

therefore $\quad 2I_{\text{diam}} = I_z = \frac{1}{2}MR^2$

Thus $\quad I_{\text{diam}} = \frac{1}{4}MR^2$

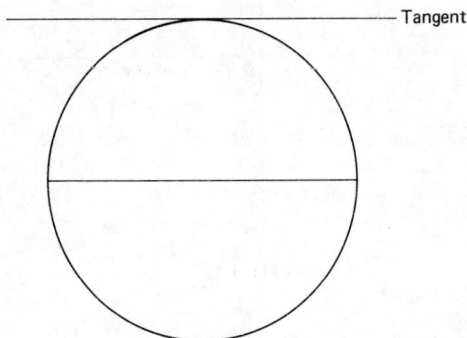

Fig. 1.14

Referring to Fig. 1.14, we see that

$$I_{\text{tangent}} = \tfrac{1}{4}MR^2 + MR^2 = 5MR^2/4$$

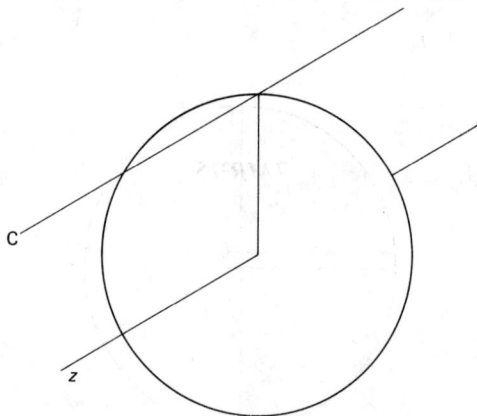

Fig. 1.15

Consider now the cylinder shown in Fig. 1.15. It is clear that

$$I_c = I_z + MR^2 = \tfrac{1}{2}MR^2 + MR^2 = 3MR^2/2$$

(b) The perpendicular axes theorem for solid figures can be written (see Fig. 1.16).

$$I_x + I_y + I_z = 2I_G$$

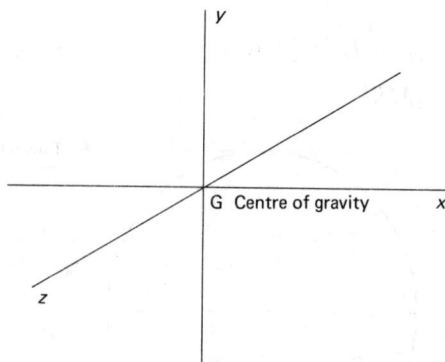

Fig. 1.16

Application to a sphere
In Section 1.4 we saw that the moment of inertia about the centre of a sphere is $3MR^2/5(\equiv I_G)$. Using form (b) of the per-

pendicular axes theorem we have

$$3I_{\text{diam}} = 2I_G$$

Therefore

$$I_{\text{diam}} = 2MR^2/5$$

From the parallel axes theorem we can write (see Fig. 1.17)

$$I_{\text{tan}} = I_{\text{diam}} + MR^2 = 7MR^2/5$$

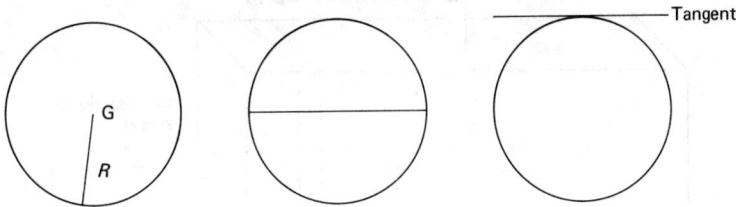

Fig. 1.17

1.7 Further calculation of moments of inertia

Cylinder
With reference to Fig. 1.18, to find I_y we use the parallel axes theorem on an element of the cylinder and then integrate:

$$dI_y = \tfrac{1}{4}(\rho\pi R^2 dx)R^2 + (\rho\pi R^2 dx)x^2$$

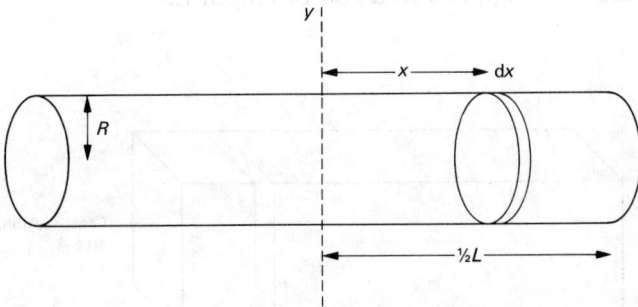

Fig. 1.18

Thus

$$I_y = 2\int_0^{L/2} \tfrac{1}{4}(\rho\pi R^2 dx)R^2 + 2\int_0^{L/2} \rho\pi R^2 x^2 dx$$

$$= 2\left[\tfrac{1}{4}\rho\pi R^4 x\right]_0^{L/2} + 2\left[\rho\pi R^2 x^3/3\right]_0^{L/2}$$

$$= \tfrac{1}{4}\rho\pi R^4 L + \rho\pi R^2 L^3/12 = M(\tfrac{1}{4}R^2 + L^2/12)$$

Thin plate
From Fig. 1.19 we have

$$I_y = 2 \int_0^{L/2} \rho dx . Ax^2 = \left[2\rho Ax^3/3 \right]_0^{L/2}$$

$$= 2\rho A \, L^3/24 = ML^2/12$$

This result also applies to a rod of length L.

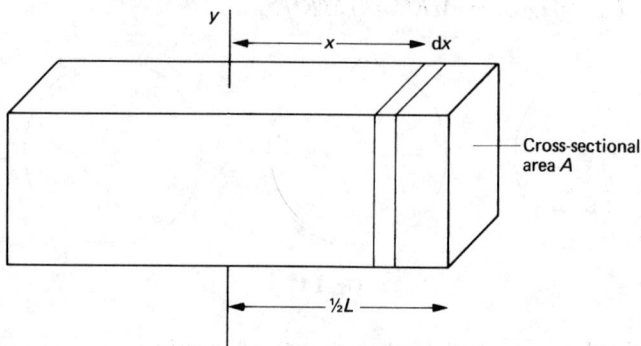

Fig. 1.19

Considering Fig. 1.20, we can write

$$I_{\text{end}} = \int_0^L \rho dx . Ax^2 = \rho \, AL^3/3 = ML^2/3$$

This result also applies to a rod of length L.

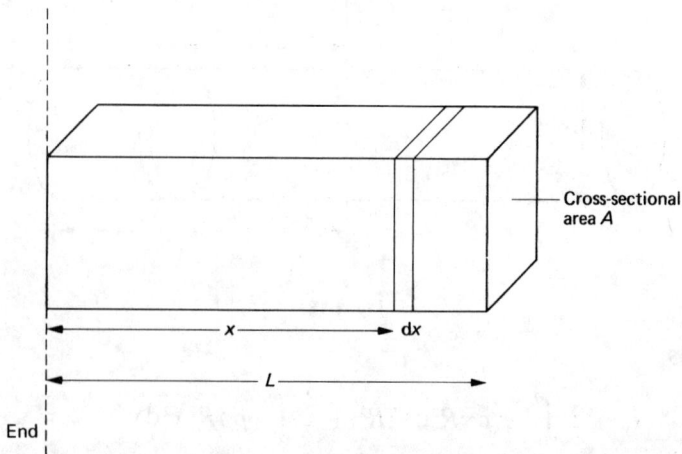

Fig. 1.20

Rectangular block

With reference to Fig. 1.21, again we use the parallel axis theorem on an element of the block before integrating:

$$dI_y = [(\rho ab dx)a^2/12] + (\rho ab dx)x^2$$

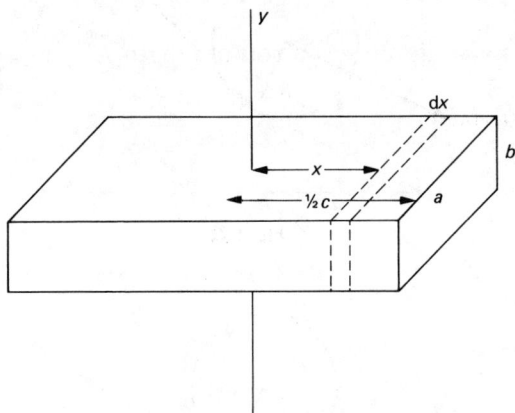

Fig. 1.21

Thus

$$I_y = 2 \int_0^{c/2} (\rho ab dx)a^2/12 + 2 \int_0^{c/2} (\rho ab dx)x^2$$
$$= [2\rho ab.a^2 x/12]_0^{c/2} + [2\rho abx^3/3]_0^{c/2}$$
$$= (\rho ab.a^2 c/12) + (\rho abc^3/12)$$
$$= M\left(\frac{a^2}{12} + \frac{c^2}{12}\right) = M\left(\frac{a^2 + c^2}{12}\right)$$

1.8 Problem of the rolling cylinder

Consider a cylinder of mass M rolling down an inclined plane with no sliding, as shown in Fig. 1.22. The motion is equivalent to rotation about an axis C, but the axis of rotation moves along with the cylinder. The angular acceleration is caused by some couple:

$$\text{couple} = Mg \sin \alpha.R$$
$$= \text{moment of inertia} \times \text{angular acceleration}$$
$$= I\ddot{\theta}$$

where g is the acceleration due to gravity, i.e.

$$Mg \sin \alpha . R = I_c \ddot{\theta}$$

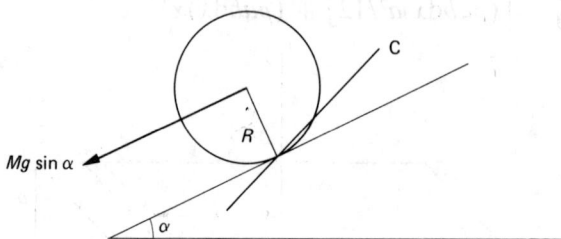

Fig. 1.22

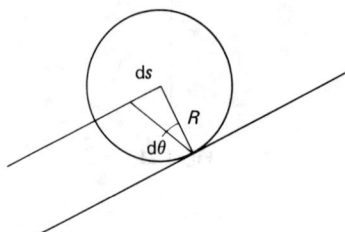

Fig. 1.23

In Fig. 1.23 ds is a small element of linear travel:

$$\mathrm{d}s = R\mathrm{d}\theta$$

so $\ddot{s} = R\ddot{\theta}$ or $\ddot{\theta} = \ddot{s}/R$

Therefore

$$RMg \sin \alpha = I_c \ddot{s}/R$$

and $\ddot{s} = (R^2 Mg \sin \alpha)/I_c$

We now introduce the radius of gyration k defined as

$$k^2 = \frac{\Sigma mr^2}{\Sigma m} = \frac{\text{moment of inertia}}{\text{mass } M}$$

By general agreement k^2 is used only when the moment of inertia is about an axis through the centre of gravity. Thus, using the moment of inertia about a parallel axis through the centre of gravity and using k, $I_c = Mk^2 + MR^2$. Therefore

$$\ddot{s} = \frac{R^2Mg \sin \alpha}{Mk^2 + MR^2} = \frac{g \sin \alpha}{1 + k^2/R^2}$$

This result also applies to a rolling sphere.

For a cylinder $k^2 = \frac{1}{2}R^2$ and $\ddot{s} = 2g \sin \alpha/3$. The time taken for the cylinder to roll down the plane is obtained by integrating this equation. Thus

$$\dot{s} = \frac{2}{3}g \sin \alpha t + C_1.$$

If we assume that the cylinder starts from rest at $t = 0$, the constant of integration C_1 will be zero.

Integrating a second time gives

$$s = (t^2g \sin \alpha/3) + C_2$$

where $C_2 = 0$ from the boundary conditions and $t = \sqrt{3s/g \sin \alpha}$.

1.9 Pendulum oscillations (compound pendulum)

With reference to Fig. 1.24 the centre of gravity is vertically below the point where the axis of suspension passes through the

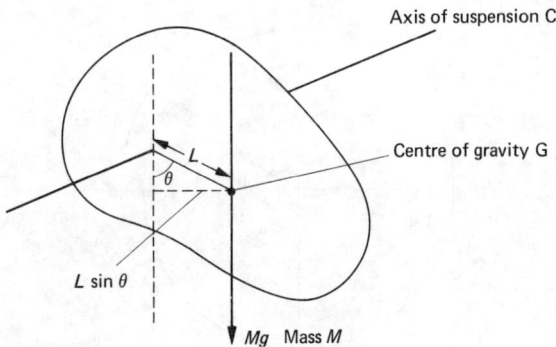

Fig. 1.24

body in the equilibrium position. When the body is deflected through an angle θ the opposing couple due to gravity is $-MgL \sin \theta = I_c\ddot{\theta}$. This is a differential equation which can only be dealt with simply if θ is small enough to approximate to sin θ. If θ is small and sin $\theta \to \theta$ radians, then

$$I_c\ddot{\theta} = -MgL\theta$$

$$\ddot{\theta} = -\frac{MgL\theta}{I_c}$$

This is the differential equation of simple harmonic motion, the solution being

$$\theta = \theta_0 \begin{cases} \sin(\omega t + \delta) \\ \cos(\omega t + \delta) \end{cases}$$

The function shown in Fig. 1.25 represents the projection on either the x- or y-axis.

The phase constant δ is determined by putting $t = 0$. If

$$\delta = 0 \quad \theta = \theta_0 \begin{cases} \sin \omega t \\ \cos \omega t \end{cases}$$

Differentiating this equation:

$$\dot{\theta} = \theta_0\omega \cos \omega t$$

Differentiating again:

$$\ddot{\theta} = -\theta_0\omega^2 \sin \omega t$$
$$= -\omega^2\theta$$

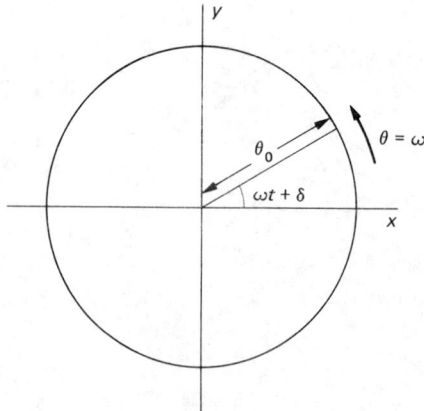

Fig. 1.25

The period is given by $T = 2\pi/\omega$. Going back to our equation

$$\ddot{\theta} = -\frac{MgL\theta}{I_c}$$

we see that the period of oscillation

$$T = 2\pi\sqrt{\frac{I_c}{MgL}}$$

But $I_c = Mk^2 + ML^2$, where Mk^2 is the moment of inertia about a parallel axis through the centre of gravity and k is the radius of gyration. Therefore

$$T = 2\pi\sqrt{\frac{Mk^2 + ML^2}{MgL}} = 2\pi\sqrt{\frac{k^2 + L^2}{gL}}$$

This is the period of oscillation of the rigid or compound pendulum. If $k^2 = 0$ then we have a simple pendulum (weightless string and mass concentrated at the centre of gravity). In this case

$$T = 2\pi\sqrt{\frac{L}{g}}$$

Properties of the compound pendulum
From Fig. 1.26 we see that

$$T = 2\pi\sqrt{\frac{k^2 + L^2}{gL}}$$

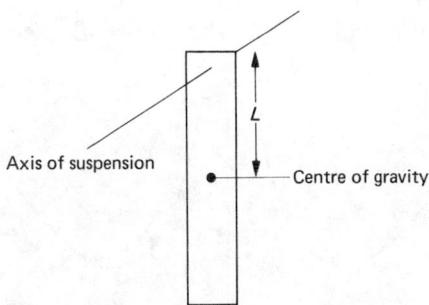

Fig. 1.26

where L is the distance between the centre of suspension and the centre of gravity. Squaring this expression we have

$$T^2 = 4\pi^2\left(\frac{k^2 + L^2}{gL}\right)$$

If we let

$$\frac{gT^2}{4\pi^2} = \frac{k^2 + L^2}{L} = \beta$$

we find that

$$L^2 - \beta L + k^2 = 0$$

which is a quadratic equation in L. Therefore

$$L = \tfrac{1}{2}\beta \pm \sqrt{\tfrac{1}{4}\beta^2 - k^2}$$

Thus we see there are two values of L (L_1 and L_2) for a given β, i.e. for a given period (see Fig. 1.27). Also, since $\sqrt{\tfrac{1}{4}\beta^2 - k^2}$ is less than $\tfrac{1}{2}\beta$ the two values of L are positive, i.e. the two points of suspension are on the same side of the centre of gravity. Returning to the quadratic in L, the product of the roots $L_1 L_2 = k^2$, and the sum of the roots $L_1 + L_2 = \beta$. Thus

$$\frac{gT^2}{4\pi^2} = L_1 + L_2$$

and

$$T = 2\pi\sqrt{\frac{L_1 + L_2}{g}}$$

T has a minimum value $T_{\min}$ when $L_1 = L_2$ (see Fig. 1.27), i.e.

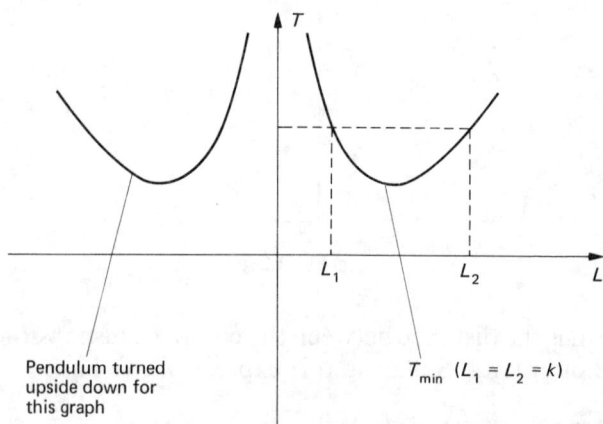

Pendulum turned
upside down for
this graph

$T_{\min}$ ($L_1 = L_2 = k$)

Fig. 1.27

when the roots are equal. But since

$$L = \tfrac{1}{2}\beta \pm \sqrt{\tfrac{1}{4}\beta^2 - k^2}$$

the condition for equal roots means $\tfrac{1}{4}\beta^2 = k^2$, i.e. $\beta_{min}^2 = 4k^2$ and $\beta_{min} = 2k$ or

$$\frac{gT_{min}^2}{4\pi^2} = 2k$$

Thus, since the sum of the roots is β, T_{min} occurs when $L_1 = L_2 = k$.

Alternatively T_{min} can be found by differentiation:

$$\frac{gT^2}{4\pi^2} = \beta = \frac{k^2 + L^2}{L}$$

Thus

$$\frac{d\beta}{dL} = \frac{2L^2 - (k^2 + L^2)}{L^2} = \frac{L^2 - k^2}{L^2}$$

To obtain the minimum value of T we put $d\beta/dL = 0$. Thus, for T_{min}

$$\frac{L^2 - k^2}{L^2} = 0$$

or

$$L^2 = k^2$$

Therefore

$$\beta_{min} = 2k^2/k = 2k$$

or

$$\frac{gT_{min}^2}{4\pi^2} = 2k$$

Equivalent simple pendulum

For the compound pendulum

$$T = 2\pi\sqrt{\frac{k^2 + L^2}{gL}}$$

In the case of a simple pendulum of length l the period is $2\pi\sqrt{l/g}$. Hence $(k^2 + L^2)/L$ corresponds to l and a simple pendulum of length $(k^2 + L^2)/L$ would have the same period as the compound pendulum. $(k^2 + L^2)/L$ is called the length of the equivalent simple pendulum.

Reversible Pendulum
With reference to Fig. 1.28 if points of suspension of equal T can be found on opposite sides of the centre of gravity, $L_1 + L_2$ yields g without a knowledge of the position of the centre of gravity since

$$T = 2\pi\sqrt{\frac{L_1 + L_2}{g}}$$

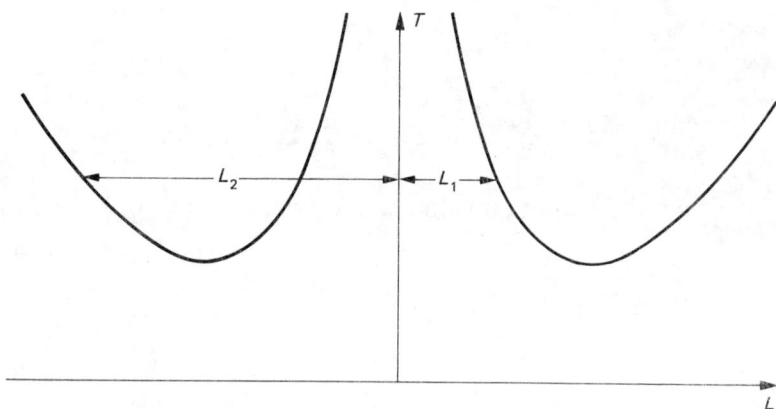

Fig. 1.28

This gives a good method of measuring g since if g is measured using a rigid pendulum $(T = 2\pi\sqrt{(k^2 + L^2)/gL})$, accuracy is limited by uncertainty in the exact position of the centre of gravity.

However, points of suspension of equal T must be found by trial and error and it is a difficult task to get two periods exactly the same. Also the pendulum must be allowed to swing for a long time in order to measure T accurately.

Kater's pendulum (measurement of g)
Here g is determined by finding two axes of suspension where the periods are nearly the same. A knowledge of the centre of gravity is required but errors in this are less important in this case.

Centre of gravity

Fig. 1.29

Referring to Fig. 1.29, points of suspension are obtained on opposite sides of the centre of gravity, period T_1 on one side, period T_2 on the other side, such that T_1 and T_2 are nearly equal. Then we have

$$\frac{gT_1^2}{4\pi^2} = \frac{k^2 + L_1^2}{L_1}$$

$$\frac{gT_2^2}{4\pi^2} = \frac{k^2 + L_2^2}{L_2}$$

Subtracting these simultaneous equations gives

$$\frac{g}{4\pi^2}(T_1^2 L_1 - T_2^2 L_2) = L_1^2 - L_2^2$$

Let

$$T_1^2 + T_2^2 = \Sigma T^2 \qquad T_1^2 - T_2^2 = \Delta T^2$$
$$L_1 + L_2 = \Sigma L \qquad L_1 - L_2 = \Delta L$$

Then

$$L_1 = \tfrac{1}{2}(\Sigma L + \Delta L) \quad L_2 = \tfrac{1}{2}(\Sigma L - \Delta L)$$

and substituting these values for L_1 and L_2 we have

$$\frac{g}{8\pi^2}\left[T_1^2(\Sigma L + \Delta L) - T_2^2(\Sigma L - \Delta L)\right] = \Sigma L.\Delta L$$

Dividing by $\Sigma L.\Delta L$ gives

$$\frac{\Sigma L(T_1^2 - T_2^2) + \Delta L(T_1^2 + T_2^2)}{\Sigma L.\Delta L} = \frac{8\pi^2}{g}$$

i.e.

$$\frac{8\pi^2}{g} = \frac{\dot{\Sigma} T^2}{\Sigma L} + \frac{\Delta T^2}{\Delta L}$$

$L_1 + L_2$ can be measured to a high degree of accuracy. The term $\Delta T^2/\Delta L$ is small. Therefore errors which arise from requiring to know the position of the centre of gravity are not so important. Clarke did this experiment at the National Physical Laboratory in Teddington, Middlesex. T_1 and T_2 were measured to seven places of decimals, ΣT^2 to six places, ΔT^2 to six places, ΣL to five places, and ΔL to two places. Moreover, $\Sigma T^2 \gg \Delta T^2$ and $\Delta L \gg \Delta T$. ΣL was obtained by an interference method. Two parallel plates were mounted on the pendulum knife edge supports and their distance apart was measured using a spectral line and an interferometer. To avoid errors which arise due to the presence of air the experiment was done in a vacuum. The final value obtained for g was $981 \cdot 1813$ cm s^{-2}.

SOURCES OF ERROR IN THE EXPERIMENT

(1) A uniform temperature is required otherwise uneven expansions occur. Temperature introduced an error in g of $\pm 0 \cdot 6$ ppm (parts per million).

(2) The amplitude of oscillation must be small since sin θ is replaced by θ to obtain the simple harmonic motion formula which is used in the analysis. Sin θ was written as an infinite series and the first few terms used. The size of the amplitude of oscillation ($\frac{3}{4}\frac{1}{4}°$) introduced an error in g of $\pm 0 \cdot 3$ ppm.

(3) Clock errors: $\pm 0 \cdot 3$ ppm.

(4) Timing errors: $\pm 1 \cdot 1$ ppm.

(5) The effect of the knife edge not being sharp was almost zero. This error was negligible.

(6) Errors due to the residual air of a high vacuum were: buoyancy error (weight of the pendulum is less due to Archimedes' principle); error due to friction as the pendulum swung through the air. In addition metals attract and cause one or two layers of molecules of air to cling to them. These three errors were negligible in high vacuum.

(7) Error due to the elasticity of the support: $+1 \cdot 5$ ppm.

(8) Error due to the elasticity of the pendulum (its own weight stretches it): $-0 \cdot 7$ ppm.

2 Gravitation

Galileo established that the acceleration f of a falling body is independent of its mass m. Some time later Newton connected pure motion with the mass moving and defined force P. Newton's second law states that

$$P = mf$$

Force is then proportional to the mass m falling. Also from Newton's third law: 'To every action there is an equal and opposite reaction', we have

force $\propto$ mass of the earth M

Thus

Force $\propto mM$

This leads to Newton's inverse square law of attraction: 'Every particle of matter attracts every other particle with a force which is proportional to the product of their masses and inversely to the square of the distance between them'.

Newton obtained this law from the results of the astronomer Kepler who had measured the orbits of some of the planets.

2.1 Kepler's laws

Kepler's first law states that: 'Each planet revolves round the sun in an ellipse with the sun at one of the foci of the ellipse' (see Fig. 2.1).

Kepler's second law states that: 'The radius vector joining the focus of the ellipse to the planet traces out equal areas in equal intervals of time'.

Newton derived this law from his system of mechanics, while

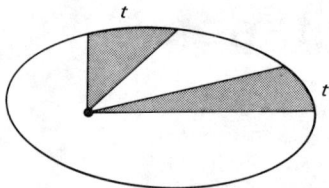

Fig. 2.1

Kepler obtained it as a generalization of Tycho Brahe's observations.

Kepler's third law can be written

$$(\text{period})^2 \propto (\text{major axis of the ellipse})^3$$

It can be proved that the first and third laws mean that each planet is experiencing an acceleration towards the sun proportional to $1/(\text{radius})^2$ for circular orbits. The second law means that this is a central acceleration (i.e. it acts along the line connecting planet and sun).

Consider circular orbits. From Kepler's second law we can write (see Fig. 2.2)

$$\frac{d}{dt}(\tfrac{1}{2}r ds) = \text{constant}$$

or

$$r ds/dt = \text{constant}$$

For a circle the radius r is constant so we have $ds/dt = $ constant. Thus, since there is a constant velocity at right angles to the line joining the masses, there is no acceleration normal to the line joining the masses.

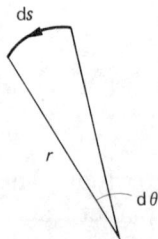

Fig. 2.2

From Kepler's first law we have

central acceleration $= r(\dot\theta)^2$

where $\dot\theta = \mathrm{d}\theta/\mathrm{d}t$ and t denotes time.
From Kepler's third law

period, $T \propto 1/\dot\theta$

Therefore

$(1/\dot\theta)^2 \propto r^3$

$\dot\theta^2 \propto 1/r^3 \quad r\dot\theta^2 \propto 1/r^2$

Thus we see that the acceleration towards the centre is proportional to $1/r^2$.

Now, Newton's law of gravitation may be written

$$F = \frac{GmM}{r^2}$$

where F is the force, m the mass of the planet, M the mass of the sun, r the distance between them, and G the universal constant of gravitation (Fig. 2.3).

Equating the force on the planet GmM/r^2 to mv^2/r, where v is the tangential velocity of the planet, we have

$$\frac{GmM}{r^2} = \frac{mv^2}{r} = mr\dot\theta^2$$

or

$$\frac{GmM}{r^2} = mr\,\frac{4\pi^2}{T^2}$$

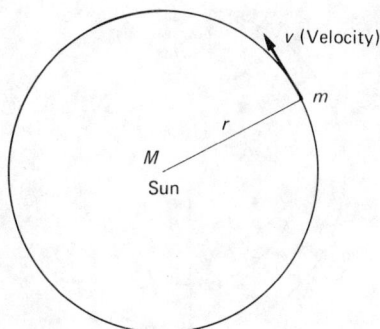

Fig. 2.3

since $T = 2\pi/\theta$. Therefore

$$\frac{GM}{4\pi^2} = \frac{r^3}{T^2} \quad \text{or} \quad T^2 = \frac{4\pi^2 r^3}{GM}$$

which also proves Kepler's third law.

2.2   Gravitation on earth

We shall use the approximation that the earth is a sphere of nearly uniform density and that gravitation outside or near the sphere is the same as if its mass were concentrated at the centre. g is called the acceleration due to gravity (also the force per unit mass or the gravitational field or the gravitational intensity).

From Newton's law of gravitation we have

$$g = \frac{GM}{R^2}$$

where M is the mass of the earth and R its radius.

But $M = \frac{4}{3}\pi R^3 \rho$, where ρ is the average density of the earth. Therefore

$$g = \frac{4}{3}\pi R \rho G$$

Variation in g with distance below the surface of the earth
Consider a pit of depth h cut through the earth's crust (Fig. 2.4). We can write

$$g_{\text{surface}} = \frac{GM}{R^2} = \frac{4}{3}\pi R \rho_{av} G$$

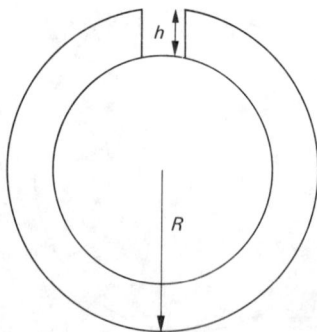

Fig. 2.4

Now, using the fact that the gravitational field inside a uniform hollow sphere is zero, the part of the earth between depth h and the surface does not contribute anything to the gravitational field. Therefore

$$g_{\text{bottom}} = \frac{GM'}{(R - h)^2}$$

where M' is the mass of a sphere of radius $R - h$:

$$M' = \tfrac{4}{3}\pi R^3 \rho_{\text{av}} - \tfrac{4}{3}\pi \left[R^3 - (R - h)^3\right] \rho_{\text{sh}}$$

where ρ_{sh} is the average density of the shell of thickness h. It is possible to find the density of the shell by measuring g with a pendulum at the surface and at the bottom of the pit. The pendulum loses time at the bottom since $g_{\text{bottom}} < g_{\text{surface}}$. Alternatively, if the density of the earth's crust ρ_{crust} is known, e.g. one knows the average density of rocks etc., then it is possible to obtain the mean density of the earth ρ_{av}. For points A and B in Fig. 2.5 we have

$$g_{\text{A}} = \frac{G}{R^2} (M_{\text{crust}} + M_{\text{core}})$$

$$= \frac{G}{R^2} \left[(4\pi R^2 h\, \rho_{\text{crust}} + \tfrac{4}{3}\pi(R - h)^3\, \rho_{\text{av}})\right]$$

$$g_{\text{B}} = \tfrac{4}{3}\pi\, \rho_{\text{av}}\, G(R - h)$$

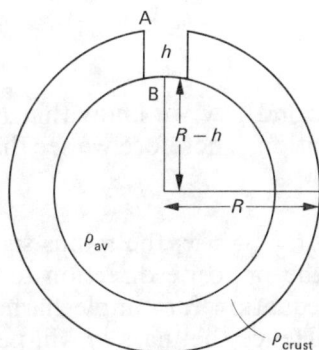

Fig. 2.5

We thus have an expression for $g_{\text{A}}/g_{\text{B}}$ in which terms in h^2 can be neglected since they are small compared with R^2. $g_{\text{A}}/g_{\text{B}}$ is then measured and gives the historical method for obtaining ρ_{av}.

With a value for ρ_{av}, since $g = \frac{4}{3}\pi\,\rho_{av}\,GR$, the value obtained for G was 64×10^{-12} kg^{-1} m^{3}s^{-2}. This compares with the up-to-date value of $66\cdot7 \times 10^{-12}$ kg^{-1} m^3 s^{-2}.

The problem of the frictionless tunnel
Consider a mass m in the tunnel at a distance x from its equilibrium position as shown in Fig. 2.6. The gravitational force acting towards centre of the earth is $\frac{4}{3}\pi r\rho Gm$, where ρ is the density of the earth. The component of this force acting along the line of the tunnel is

$$\frac{4}{3}\pi r\rho Gm \cos \alpha = \frac{4}{3}\pi r\rho Gm \frac{x}{r}$$

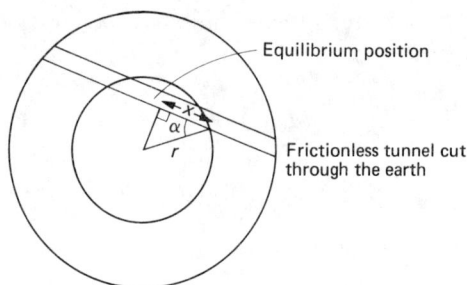

Fig. 2.6

From Newton's second law we know that $P = mf$ where P is force and f acceleration. Therefore we see that

$$f = \frac{4}{3}\pi\rho x G$$

This is of the form $f = -\omega^2 x$; the minus sign occurs since the acceleration is in the opposite direction to x increasing. But $f = -\omega^2 x$ is the equation for simple harmonic motion with period $T = 2\pi/\omega$. Hence the mass m will perform simple harmonic motion in the tunnel with period $T = 2\pi\sqrt{3/4G\pi\rho}$. Thus the period is independent of where the tunnel is cut.

Initial velocity required to escape from earth's gravitational field
We shall consider a particle of mass m projected vertically up-

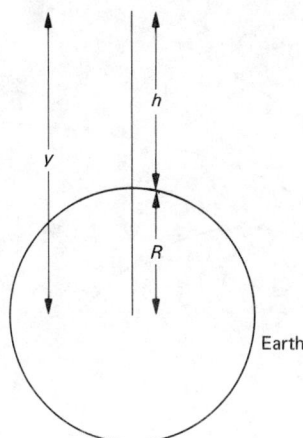

Fig. 2.7

wards with an initial velocity v_0 (see Fig. 2.7) and we shall use the principle of conservation of energy. The initial kinetic energy of the particle is $\frac{1}{2}mv_0^2$. At any time after launch the velocity is reduced to some value v at some distance y from the centre of the earth. The kinetic energy is then $\frac{1}{2}mv^2$. Thus the loss of kinetic energy is $\frac{1}{2}m(v_0^2 - v^2)$. The work done is given by the product of force and distance moved. Thus the total work done is given by

$$\int_R^{R+h} \frac{GmM}{y^2} \, dy$$

where M is the mass of the earth. Integrating this expression we find that the work done is

$$-\left[\frac{GmM}{y}\right]_R^{R+h} = \frac{GmM}{R} - \frac{GmM}{R+h}$$

We can equate the loss of kinetic energy with this work done:

$$v^2 = v_0^2 - \frac{2GM}{R} + \frac{2GM}{R+h}$$

As h increases the last term gets smaller. There are three cases to consider: $v_0^2 < 2GM/R$, then v reaches zero and the body will fall back to earth; $v_0^2 > 2GM/R$, now $v^2 \to$ constant $= v_0^2 - 2GM/R$; $v_0^2 = 2GM/R$, now $v^2 \to 0$.

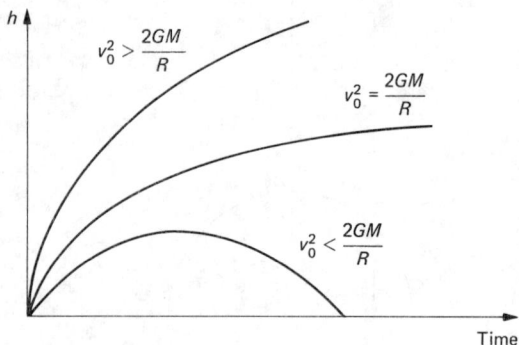

Fig. 2.8

The escape velocity is given by (see Fig. 2.8)

$$v_0^2 = \frac{2GM}{R} = 2gR$$

since $g = MG/R^2$.

Variation of g with latitude
There is a slight change in g both in magnitude and direction due to the earth's spin. The effect is zero at the poles and a maximum at the equator. With reference to Fig. 2.9, the main gravitational force is directed towards the centre of the earth and this maintains the acceleration towards the centre of the circle of radius $R \cos L$ as well as giving weight effects. The main acceleration

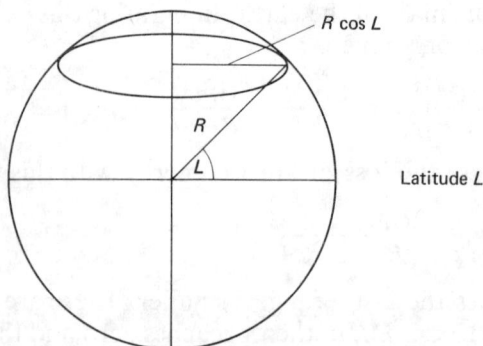

Fig. 2.9

$p = GM/R^2$. The acceleration due to the earth's spin, c, is given by

$$c = r\dot{\theta}^2 = R \cos L \; \dot{\theta}^2$$

where $\dot{\theta} = 2\pi/$(period of earth's spin). Therefore

$$c = R \cos L \frac{4\pi^2}{T^2}$$

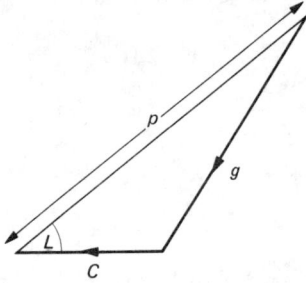

Fig. 2.10

But, from Fig. 2.10, we have

$$g^2 = p^2 + c^2 - 2pc \cos L$$

Since $c \ll p$ we can neglect c^2. Thus

$$g^2 = p^2 \left(1 - 2\frac{c}{p} \cos L\right)$$

and

$$g = p \left(1 - 2\frac{c}{p} \cos L\right)^{1/2}$$

We can now expand the right-hand side as a series using the binomial theorem and retain only the first-order term since $c \ll p$. Hence

$$g = p \left(1 - \frac{c}{p} \cos L\right) = p - c \cos L$$

From Fig. 2.11 we see that

$$\sin \theta = c \sin L/g$$

At the pole $L = 90°$ and $\cos L = 0$. Thus the acceleration due to gravity at the pole is p. At the equator $L = 0°$ and $\cos L = 1$.

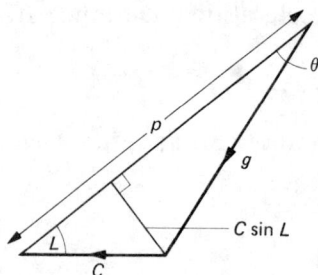

Fig. 2.11

Therefore, at the equator

$$g = p - c = p - \frac{4\pi^2}{T^2} R$$

Variations in g due to latitude are quite large (9·78–9·81 m s^{-2}) but there are other effects.

Acceleration due to gravity slightly above the earth (e.g. on mountains)
Consider the two points A and B shown in Fig. 2.12:

$$g_A = \frac{GM}{R^2} \quad g_B = \frac{GM}{(R + h)^2}$$

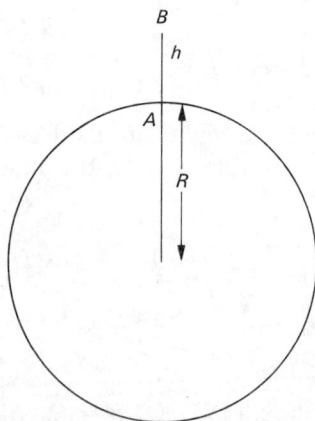

Fig. 2.12

The difference in accelerations is

$$\Delta g = \frac{GM}{R^2} - \frac{GM}{(R + h)^2}$$

However, for small changes it is better to use the calculus.

$$\Delta g = - \frac{2GM}{R^3} \Delta R = -2g \frac{\Delta R}{R}$$

$$\frac{\Delta g}{g} = -2 \frac{\Delta R}{R}$$

This treatment neglects the gravitational effect of the mountain itself.

2.3 Measurement of G

Since we are dealing with a very small force its effect between two objects is difficult to detect. For accurate results high sensitivity or large masses are required. Field determinations are limited in accuracy due to the density problem, e.g. a mountain on a plane might be used, but there will be uncertainty in the value of the density of the mountain.

Boys' Experiment
Boys' experiment is a modification of the original Cavendish experiment performed at Cambridge. The experiment uses the attraction between two lead balls, each of approximately 7 kg mass, and two gold balls, each of approximately 3 g mass. A diagram of the apparatus is shown in Fig. 2.13. The mirror B, supported by a quartz fibre, serves as a torsion arm; the position and movement of the small spheres can be observed by means of a beam of light reflected by the mirror. The equilibrium position is controlled by the torsion head.

The high accuracy of this method is due to the fineness and good elastic properties of the suspension, the reduction in temperature disturbance because of the smallness of the space in which the suspension system is situated; also, the large spheres are positioned so as to exert a maximum couple on the suspended system.

We now give the theory behind the experiment (see Fig. 2.14). The two small masses are attracted by the two large masses, the

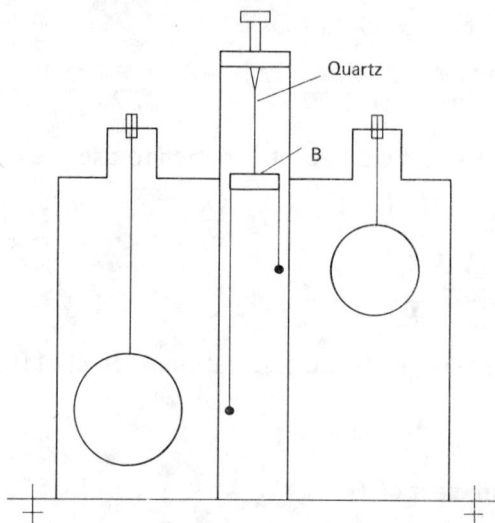

Fig. 2.13

deflection being θ. At equilibrium the couple tending to rotate the small masses is balanced by the torsional resistance of the suspension given by $\theta \times C$, where C is the torsional constant of the suspension. But the couple acting is given by

$$\frac{GMmp}{d^2}$$

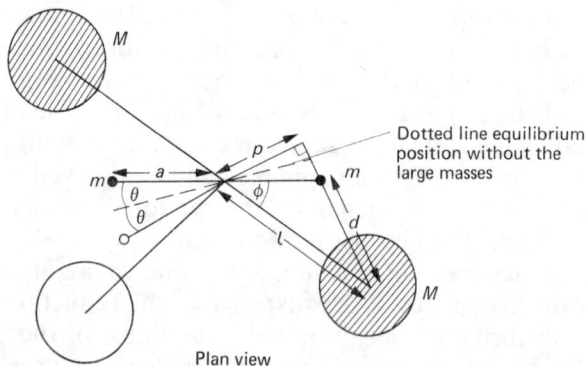

Dotted line equilibrium position without the large masses

Plan view

Fig. 2.14

Now we require to express p and d^2 in terms of a, l and ϕ (see Fig. 2.14), which can be measured accurately. From Fig. 2.15

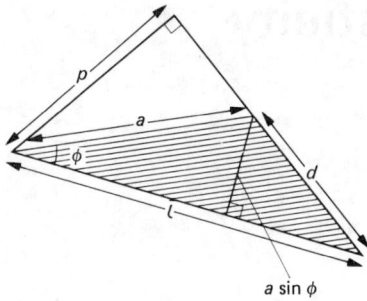

Fig. 2.15

we have

$$d^2 = a^2 + l^2 - 2al \cos \phi$$

The area of the shaded triangle is $\frac{1}{2}pd = \frac{1}{2}la \sin \phi$, therefore

$$p = la \sin \phi / d$$

and the couple can be written

$$\frac{GmMla \sin \phi}{d^3} = \frac{GmMla \sin \phi}{(a^2 + l^2 - 2al \cos \phi)^{3/2}} = C\theta$$

Everything in this equation can be measured except G and so G is obtained. The torsional constant C is determined by measuring the period of the suspended system without the large masses. In this case

$$T^2 = 4\pi^2 \frac{I}{C}$$

where I is the moment of inertia about the axis of suspension. To allow for lack of symmetry the large masses are rotated through a right angle in which case the deflection change should be 2θ. The mean of the two deflections is then taken.

Boys obtained a value for G of $66 \cdot 5 \times 10^{-12}$ SI units.

Experiments show that G does not depend on the medium between the masses since G has been determined for various media. The effect of the medium seems to occur when one has a positive and negative convention, e.g. in electrostatics and magnetostatics, but the basic property of mass appears to be of one sign only. There is also no dependence on chemical composition of the masses, no dependence on direction, and no dependence on nuclear instability.

3 Elasticity

In elasticity quantities are required which depend on the material only and not on its dimensions. Stress is defined as force per unit area while the fractional deformation is the strain.

3.1 Behaviour of materials under stress

A typical stress strain curve is shown in Fig. 3.1. The features exhibited by this curve are listed as follows:

(1) Small stress. Stress is proportional to strain (Hooke's law) as far as the limit of proportionality.
(2) Larger stress. Strain is a function of stress, but not proportional to it, as far as the elastic limit.
(3) Still larger stress. Beyond the elastic limit effects are no longer reversible, and the material becomes partially

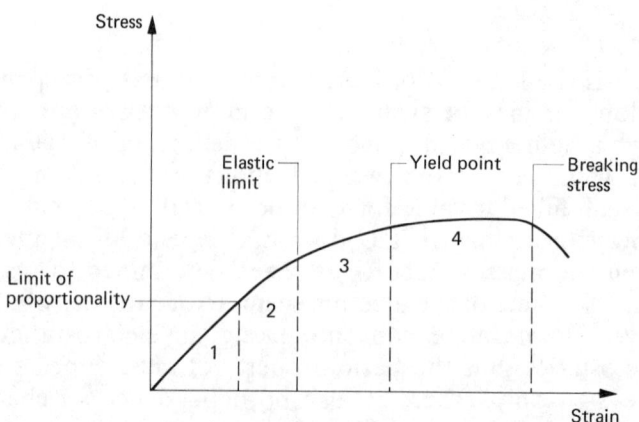

Fig. 3.1

plastic. The plastic property means that when the stress is removed some strain remains. As the stress is increased eventually the yield point is reached.

(4) Still larger stress. Beyond the yield point the material begins to flow and a large increase in strain occurs for scarcely any increase in stress. Eventually the material ceases to resist, stress is a maximum (breaking stress) and soon the material breaks.

Most materials behave as in Fig. 3.1 and the linear region is quite large for metals. Slipping of molecules or groups of molecules begins to occur at the limit of proportionality. Eventually the molecules slide freely, cracks and lines appear, weak areas conglomerate and ultimately the material breaks. For most materials the part of the curve between the limit of proportionality and the elastic limit is quite small.

Longitudinal deformation
With reference to Fig. 3.2, L is the length of the material before it is deformed. The material is stretched until the internal molecular resistance reaches a state of equilibrium with the force. We are concerned only with the Hooke's law region:

$$\frac{\text{stress}}{\text{strain}} = \text{constant} = \text{modulus}$$

$$= \frac{F/A}{dL/L} = Y$$

where Y is Young's modulus. It is found experimentally that whenever a material is stretched, there is a contraction in all directions at right angles to the stretching direction. Thus the stretching is accompanied by a lateral contraction.

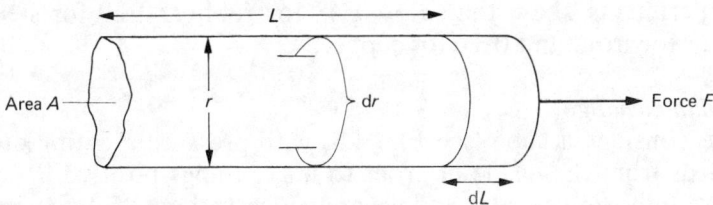

Fig. 3.2

Poisson's ratio, σ, is defined as

$$-\sigma = \frac{\mathrm{d}r/r}{\mathrm{d}L/L}$$

where r is the original diameter and $\mathrm{d}r$ is the lateral contraction. Therefore

$$-\frac{\mathrm{d}r}{r} = \sigma\frac{\mathrm{d}L}{L}$$

But

$$\frac{\mathrm{d}L}{L} = \frac{F/A}{Y}$$

therefore

$$-\frac{\mathrm{d}r}{r} = \sigma\frac{F/A}{Y}$$

When the volume of a strip of material remains constant while an extension and a lateral contraction takes place, it can easily be shown, as follows, that Poisson's ratio is 0·5 in this case. Volume $V = \pi r^2 l$, where r is the radius, and l the length of a cylinder of the material. Differentiating both sides, we have

$$0 = \pi r^2 \mathrm{d}l + 2\pi r \mathrm{d}r l$$

therefore

$$r\mathrm{d}l = -2l\mathrm{d}r$$

and so

$$-\frac{\mathrm{d}r/r}{\mathrm{d}l/l} = \frac{1}{2}$$

Experiments show that $\sigma = 0·48$ for rubber, $0·29$ for steel, $0·27$ for iron, and $0·26$ for copper.

Volume change

We consider a cube (see Fig. 3.3) with pressure P acting outwards from the inside in order to make things positive. P is a pure uniform pressure and we are concerned only with change of volume (no change of shape occurs). Now we define the

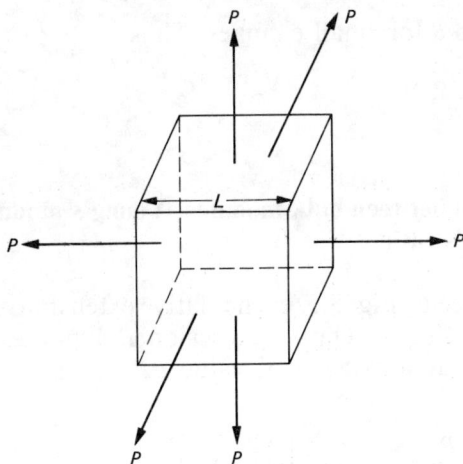

Fig. 3.3

bulk modulus of elasticity, k, as

$$k = \frac{\text{stress}}{\text{strain}} = \frac{P}{\mathrm{d}V/V}$$

where V is the volume of the cube. Therefore

$$\frac{\mathrm{d}V}{V} = \frac{P}{k}$$

Shear strain

When a rod is twisted a shear strain is produced. We shall consider the simplest situation giving rise to a shear strain.

When the base of the body is fixed a tangential force F distorts the body as shown in Fig. 3.4. The rigidity or shear modulus, n, is given by

$$n = \frac{F/A}{\theta}$$

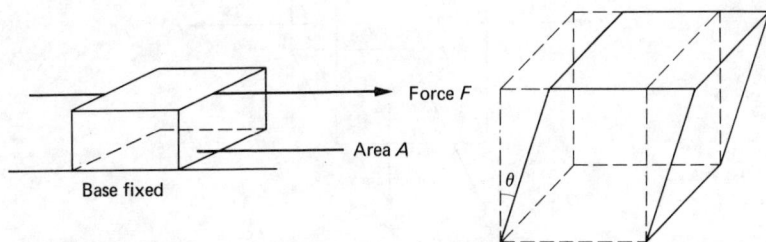

Force F

Area A

Base fixed

Fig. 3.4

since $\theta = \tan \theta$ for small changes. Thus

$$\theta = \frac{F/A}{n}$$

3.2   Relation between bulk modulus, Young's modulus and Poisson's ratio

With reference to Fig. 3.5, each relative extension dL/L is made up of a pure Young relative extension P/Y minus two Poisson relative contractions each of value $\sigma P/Y$. Thus each relative extension can be written

$$\frac{dL}{L} = \frac{P}{Y} - 2\sigma\frac{P}{Y} = \frac{P}{Y}(1 - 2\sigma)$$

We now introduce the bulk modulus. The relative volume change is

$$\frac{d(L^3)}{L^3} = \frac{3L^2 dL}{L^3} = \frac{3dL}{L} = \frac{dV}{V}$$

But $dV/V = P/k$, and so

$$\frac{P}{k} = \frac{3\,dL}{L} = \frac{3P}{Y}(1 - 2\sigma)$$

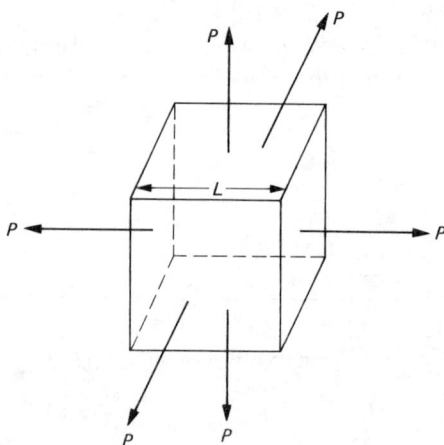

Fig. 3.5

Therefore $\quad Y = 3k(1 - 2\sigma)$

If it is difficult to measure a particular modulus it may be obtained by first determining the more easily measurable moduli and then using the above equation.

3.3 Relation between Young's modulus, rigidity modulus and Poisson's ratio

With reference to Figures 3.6 and 3.7 the forces necessary to produce a shear strain and to prevent the body from moving

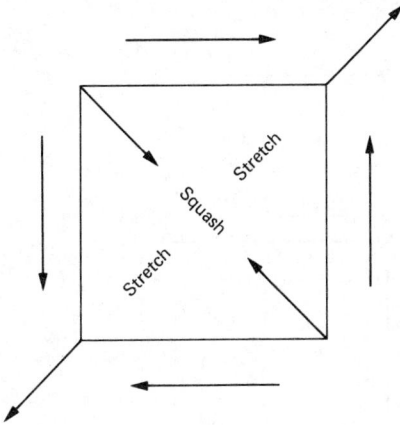

Fig. 3.6

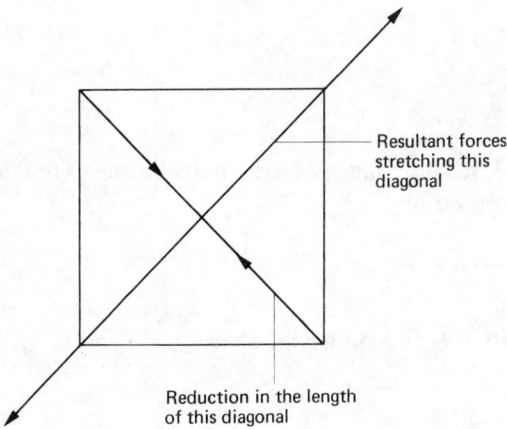

Resultant forces stretching this diagonal

Reduction in the length of this diagonal

Fig. 3.7

are shown. The shear situation amounts to an extension of one diagonal and all directions parallel to this diagonal are stretched.

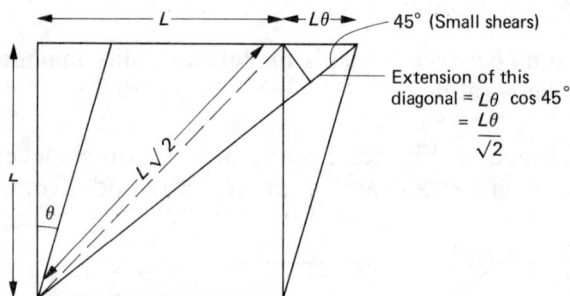

Fig. 3.8

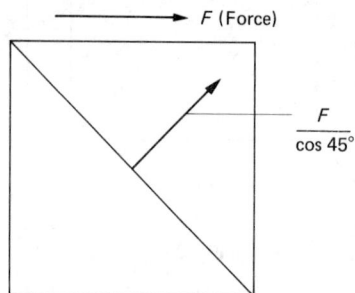

Fig. 3.9

From Fig. 3.8, the diagonal strain (relative extension of the diagonal) is given by

$$\frac{L\theta}{\sqrt{2}} \frac{1}{L\sqrt{2}} = \frac{\theta}{2}$$

and from Fig. 3.9 the diagonal stress required to produce this strain is

$$\frac{F/\cos 45}{L^2/\cos 45} = \frac{F}{L^2} = \frac{F}{A}$$

i.e. the diagonal stress equals the tangential stress, since the diagonal force $F/\cos 45$ acts on an area $A/\cos 45$.

However, the actual extension of the diagonal is assisted by a compressing force on the other diagonal. Thus the relative extension $\theta/2$ is a pure Young extension $(F/A)/Y$ plus one Poisson effect $\sigma(F/A)/Y$. Therefore

$$\frac{\theta}{2} = \frac{F/A}{Y} + \sigma\frac{F/A}{Y}$$

$$= \frac{F/A}{Y}(1 + \sigma)$$

But $(F/A)/\theta = n$, the rigidity modulus. So

$$\frac{\theta}{2} = \frac{n\theta}{Y}(1 + \sigma) \qquad \text{or} \qquad Y = 2n(1 + \sigma)$$

Thus

$$Y = 3k(1 - 2\sigma) \quad \text{and} \quad Y = 2n(1 + \sigma).$$

From these two equations the relationship between any three of the quantities Y, n, k and σ may be obtained. For example, to find the relationship between Y, n and k we can write

$$3k(1 - 2\sigma) = 2n(1 + \sigma)$$

Now k and n are always positive. If σ is positive, the left-hand side cannot be negative and so $\sigma \not> \frac{1}{2}$. If σ is negative, the right-hand side cannot be negative so that $\sigma \not< -1$. For the vast majority of materials a reduction in lateral dimension occurs and therefore σ is positive. In recent times materials have been made with a negative σ (increase in lateral dimension). Using the equation $Y = 2n(1 + \sigma)$, i.e. $\sigma = (Y/2n) - 1$, and substituting for σ in $Y = 3k(1 - 2\sigma)$, we have

$$Y = 3k\left[1 - 2\left(\frac{Y}{2n} - 1\right)\right]$$

$$= 3k - \frac{3kY}{n} + 6k$$

Therefore

$$Yn = 3kn - 3kY + 6nk$$

and so $Y(n + 3k) = 9nk$

which gives the relationship between Y, n and k.

3.4   Torsion of a cylinder

The twisting of a cylindrical wire is the basis of the torsion suspension of a galvanometer and is a shear problem. With reference to Fig. 3.10 the cylinder is twisted from the top through

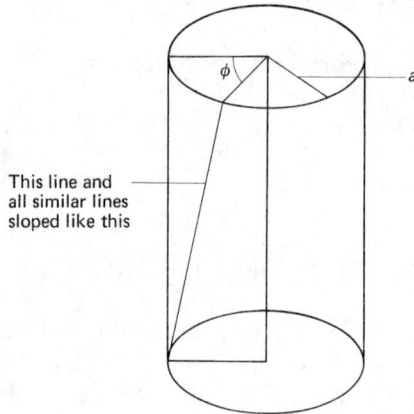

This line and all similar lines sloped like this

Fig. 3.10

an angle ϕ while the base is fixed. Consider a thin cylindrical shell at a distance r from the axis and of thickness dr as in Fig. 3.11. Fig. 3.12 shows the shell stretched out. We have

$$r\phi = L\theta$$

For this thin shell the stress (a small part of the total stress on the cylinder) causing the strain is

$$\text{element of stress} = \frac{dF}{2\pi r dr}$$

But the rigidity modulus, n, is given by

$$\frac{dF/2\pi r dr}{\theta}$$

thus $dF = 2\pi r dr n\theta$

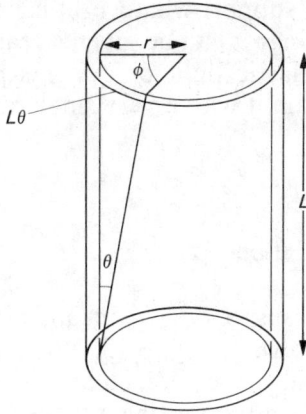

Fig. 3.11

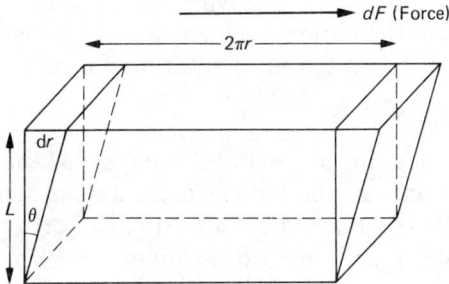

Fig. 3.12

The element of torque is given by

$$r\,\mathrm{d}F = 2\pi r^2 \mathrm{d}r n\theta = \frac{n\phi}{L} 2\pi r^3 \mathrm{d}r$$

The total torque, G say, is obtained by integration:

$$G = \frac{2\pi n\phi}{L} \int_0^a r^3 \mathrm{d}r = \frac{\pi na^4}{2L} \phi = C\phi$$

In the case of a system suspended by a wire the term $\pi na^4/2L$ $= C$, depending on the dimensions and elastic properties of the wire, is called the torsional constant of the suspension. Thus for high sensitivity (large angle of twist for a given torque) C must be small, i.e. a must be small and L large. However, if the wire is made too thin and long it takes a very long time to settle down after a torque is applied. Fused quartz is better than a metal for use as a suspension for the following reasons: it is strong and thus will support whatever it has to support without stretching or breaking; it can be given a reasonable twist without exceeding its elastic limit or even its proportional limit; finally, one can obtain it with a very small radius of remarkable uniformity.

3.5 Torsional oscillations

Consider a mass M suspended by means of a wire of radius a and length L, as shown in Fig. 3.13. When a torque is applied to the wire externally the molecular resistance of the wire provides an equal and opposing torque. Thus the external torque is opposed by an internal torque $C\phi$, where $C = \pi na^4/2L$; n is the rigidity modulus of the wire. The applied couple is given by the product of the moment of inertia and the angular acceleration. Therefore the equation of rotational motion is

$$I\ddot{\phi} = -C\phi$$

where I is the moment of inertia of the mass M about the axis of the wire. The mass of the wire is neglected as being negligible compared with the mass M. This equation represents simple harmonic motion with period T_T given by $2\pi\sqrt{I/C}$. Thus

$$\frac{T_T^2}{4\pi^2} = \frac{I}{C} = \frac{2LI}{\pi na^4}$$

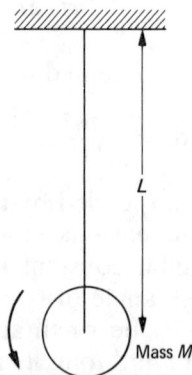

Fig. 3.13

If the axis of suspension passes through the centre of gravity of the mass M, $I = Mk^2$, where k denotes radius of gyration. Therefore

$$\frac{T_T^2}{4\pi^2} = \frac{2LMk^2}{\pi na^4}$$

and if the mass M is a disc of radius R, $k^2 = \frac{1}{2}R^2$ and so

$$\frac{T_T^2}{4\pi^2} = \frac{MR^2L}{\pi na^4}$$

3.6 Longitudinal oscillations

With reference to Fig. 3.14, consider a mass M suspended by means of a wire of length L and radius a. If y represents a small extension of the wire

$$\frac{\text{Stress}}{\text{Strain}} = \frac{Mg/\pi a^2}{y/L} = Y$$

where Y denotes Young's modulus. Therefore

$$Mg = \frac{Yy\pi a^2}{L} = \frac{Y\pi a^2 y}{L}$$

Thus force is proportional to extension.

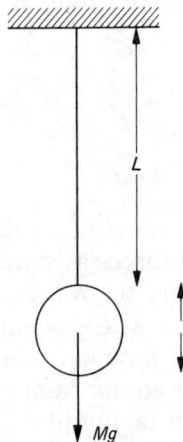

Fig. 3.14

If the mass M is displaced from its equilibrium position and then released, it returns towards its equilibrium position with an acceleration $\ddot{y}$, and if y is the extension at any instant, in general a force $Y\pi a^2 y/L$ opposes the extension y. Thus

$$M\ddot{y} = -\frac{Y\pi a^2 y}{L}$$

This equation represents simple harmonic motion with period of oscillation

$$T_{\text{L}} = 2\pi\sqrt{\frac{ML}{Y\pi a^2}}$$

and so

$$\frac{T_{\text{L}}^2}{4\pi^2} = \frac{ML}{Y\pi a^2}$$

But for torsional oscillations

$$\frac{T_{\text{T}}^2}{4\pi^2} = \frac{2LMk^2}{\pi n a^4}$$

thus

$$\frac{T_{\text{T}}^2}{T_{\text{L}}^2} = \frac{2k^2 Y\pi a^2}{\pi n a^4} = 2k^2 \frac{Y}{n}\frac{1}{a^2}$$

and if the mass M is a disc of radius R

$$\frac{T_{\text{T}}^2}{T_{\text{L}}^2} = \frac{Y}{n}\frac{R^2}{a^2}$$

3.7 Bending of a beam

We shall consider the bending of an initially horizontal beam and the action of vertical forces as shown in Fig. 3.15. External forces cause a bending and there is a resistance to this arising from the elastic properties of the beam. For any section of the beam there is a bending moment which causes the beam to distort until it comes into equilibrium with this moment. There is also a certain shearing taking place in addition to the pure bending, but this effect is small compared with the bending and may therefore be neglected.

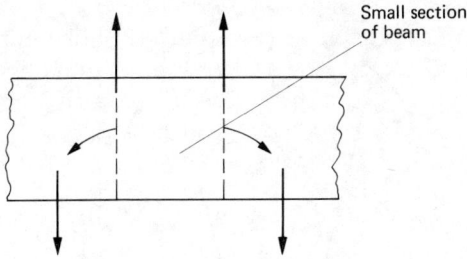

Force system in the case of a bending beam

Fig. 3.15

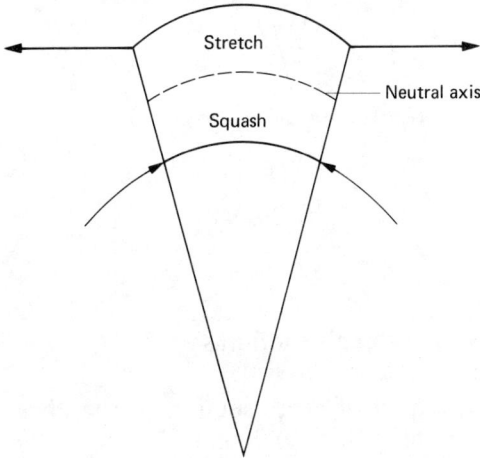

Fig. 3.16

With reference to Fig. 3.16, the moment effect produces an extension of everything above what is called the neutral surface (indicated by the dotted line and half way up for the simple shapes we shall deal with) and a compression of everything below it. A filament of the beam lying along the neutral axis remains unchanged in length. We shall use the fact that for a small arc any continuous curve approximates to the arc of a circle. Consider a fibre of the beam (Fig. 8.17) at a distance Z above the neutral surface. The length of such a fibre is $(R + Z)\phi$. The extension relative to a fibre in the neutral surface which is not stretched is $(R + Z)\phi - R\phi = Z\phi$. Therefore, for this fibre,

$$\text{strain} = \frac{Z\phi}{R\phi} = \frac{Z}{R}$$

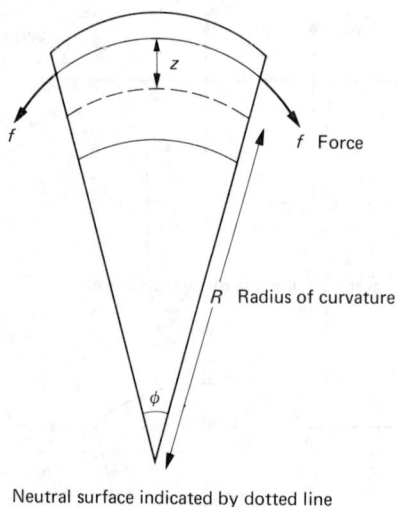

Neutral surface indicated by dotted line

Fig. 3.17

But

$$\frac{f/\mathrm{d}A}{Z/R} = Y(\text{Young's modulus})$$

where $\mathrm{d}A$ is the area of cross section of the fibre. Therefore, the force f stretching this particular fibre is

$$\frac{Y\mathrm{d}A}{R} Z$$

The moment of this force about the neutral axis is

$$fZ = \frac{Y\mathrm{d}A}{R} Z^2$$

which applies to a particular fibre. The total moment considering all fibres is given by

$$\frac{Y}{R} \Sigma \, \mathrm{d}A \, Z^2$$

Introducing the radius of gyration k defined by

$$k^2 = \frac{\Sigma \, \mathrm{d}A \, Z^2}{\Sigma \, \mathrm{d}A} = \frac{\Sigma \, \mathrm{d}A \, Z^2}{A}$$

where A is the area of cross section of the beam, the total moment is therefore

$$\frac{Y}{R}Ak^2$$

Calculation of k for a circular cross section (cylindrical rod)
In Fig. 3.18 the neutral axis is the diameter, thus

$$k_z^2 = \tfrac{1}{2}R^2 \text{ (see calculation of moments of inertia, §1.4)}$$

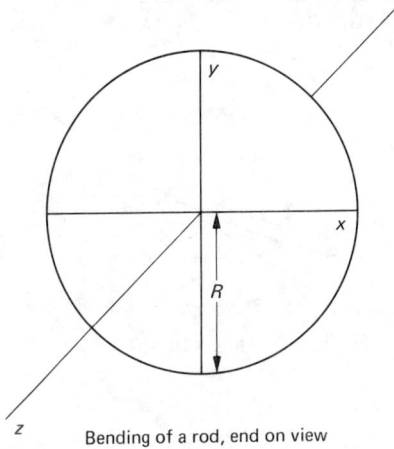

Bending of a rod, end on view

Fig. 3.18

But

$$k_z^2 = k_x^2 + k_y^2 = 2k^2$$

Therefore

$$2k^2 = \tfrac{1}{2}R^2 \quad \text{and} \quad k^2 = \tfrac{1}{4}R^2$$

If A is the area of cross section, we have

$$Ak^2 = \pi R^2 \times \tfrac{1}{4}R^2 = \tfrac{1}{4}\pi R^4$$

Calculation of k for a rectangular cross section
For a beam of rectangular cross section $k^2 = \tfrac{1}{12}b^2$ (see calculation of moments of inertia, § 1.4). Therefore from Fig. 8.19

$$Ak^2 = \tfrac{1}{12}ab^3$$

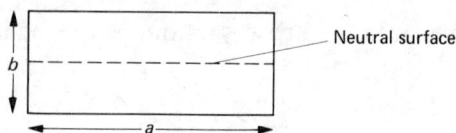

Neutral surface

Fig. 3.19

Curvature
In Fig. 3.20, Curvature K is defined as $d\psi/dS$. Thus

$$K = \frac{d\psi}{dS} = \frac{d\psi}{dx}\frac{dx}{dS} = \cos\psi\,\frac{d\psi}{dx}$$

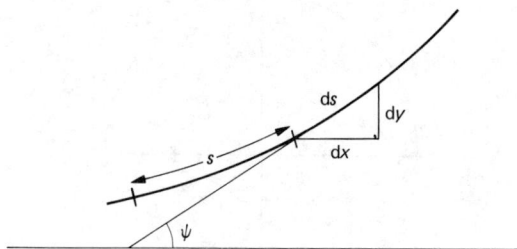

Fig. 3.20

But

$$\tan\psi = dy/dx$$

therefore

$$\psi = \tan^{-1}(dy/dx)$$

and so

$$\frac{d\psi}{dx} = \frac{1}{1+(dy/dx)^2}\frac{d^2y}{dx^2}$$

Also

$$\cos\psi = \frac{1}{\sec\psi} = \frac{1}{(1+\tan^2\psi)^{1/2}} = \frac{1}{\left[1+(dy/dx)^2\right]^{1/2}}$$

Therefore

$$K = \frac{d^2y/dx^2}{\left[1+(dy/dx)^2\right]^{3/2}}$$

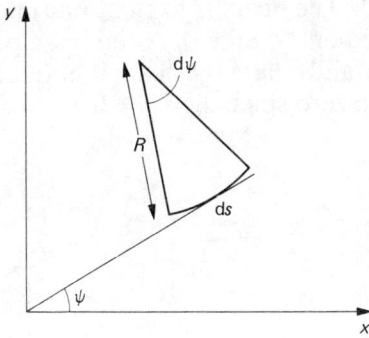

Fig. 3.21

With reference to Fig. 3.21

$$1/R = K$$
$$= \frac{d^2y/dx^2}{\left[1 + (dy/dx)^2\right]^{3/2}}$$

where R denotes radius of curvature. When dealing with the small bending of an initially horizontal beam, the approximation $1/R = d^2y/dx^2$ is used, since $(dy/dx)^2$ will be small.

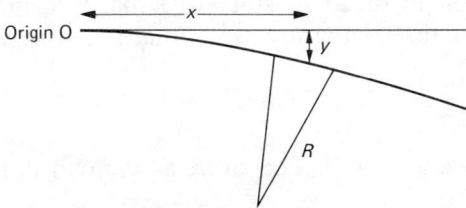

Fig. 3.22

For the small bending of an initially horizontal beam (Fig. 3.22) the total moment is

$$\frac{Y}{R} Ak^2 = YAk^2 \frac{d^2y}{dx^2} = B$$

where B is the moment due to external forces.

Light cantilever
By a light cantilever we mean that the weight of the beam is small compared with that of any object placed on it; and thus it

may be neglected. The beam is fixed at one end as in Fig. 3.23. It is always convenient to take the origin at a point where there is no depression y and where the beam is horizontal. Integrating constants are then zero since $dy/dx = 0$ when $x = 0$, and $y = 0$ when $x = 0$.

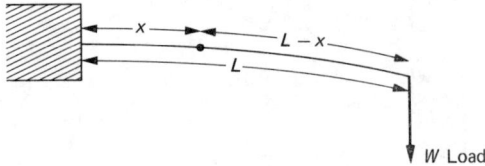

Fig. 3.23.

We have

$$YAk^2 \frac{d^2y}{dx^2} = YAk^2y''$$

This is the moment due to external forces and so

$$YAk^2y'' = W(L - x)$$

We assume that the bending is sufficiently slight for the distance from the origin in the bent state to equal that in the unbent state. We can now integrate this equation and we see that

$$YAk^2 \frac{dy}{dx} = YAk^2y' = W(Lx - \tfrac{1}{2}x^2) + 0$$

since $y' = 0$ when $x = 0$. The slope at the end of the beam y'_E is given by

$$YAk^2y'_E = \tfrac{1}{2}WL^2$$

Integrating again we have

$$YAk^2y = W(\tfrac{1}{2}Lx^2 - \tfrac{1}{6}x^3) + 0$$

since $y = 0$ when $x = 0$. Putting $x = L$ to obtain the depression at the end y_E, we see that

$$YAk^2y_E = WL^3/3$$

For some intermediate point P along the beam at $x = pL$, where p is a fraction, we have

$$YAk^2y_P = W(\tfrac{1}{2}p^2L^3 - \tfrac{1}{6}p^3L^3) = \tfrac{1}{2}WL^3(p^2 - \tfrac{1}{3}p^3)$$

LIGHT CANTILEVER (LOAD NOT AT THE END)

Consider the bending of a beam of length L (Fig. 3.24), under the action of a load W placed at a point P on the beam at a distance pL along the beam, where p is some fraction. The moment due to external forces can be put equal to YAk^2y''. Thus

$$YAk^2y'' = W(pL - x)$$

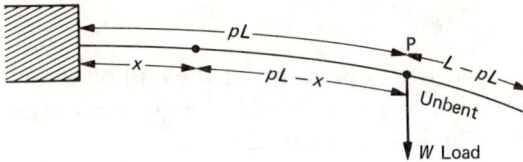

Fig. 3.24

Here care must be taken to note that this equation applies only to points between the origin and the load. Integrating this equation we have

$$YAk^2y' = W(pLx - \tfrac{1}{2}x^2) + 0$$

Thus the slope at P, y'_P is given by

$$YAk^2y'_P = W(p^2L^2 - \tfrac{1}{2}p^2L^2)$$
$$= \tfrac{1}{2}Wp^2L^2$$

Integrating again:

$$YAk^2y = W(\tfrac{1}{2}pLx^2 - \tfrac{1}{6}x^3)$$

Thus the depression at P, y_P is given by

$$YAk^2y_P = W(\tfrac{1}{2}p^3L^3 - \tfrac{1}{6}p^3L^3) = \tfrac{1}{3}Wp^3L^3$$

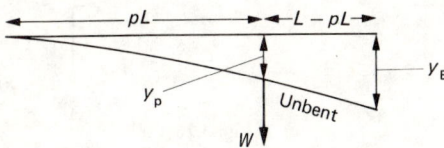

Fig. 3.25

Referring to Fig. 3.25, the depression at the end of the beam y_E, when $x = L$, is given by

$$y_E = y_P + (L - pL)y'_P$$

$$= \frac{1}{YAk^2} \left[\tfrac{1}{3}Wp^3L^3 + \tfrac{1}{2}(L - pL)Wp^2L^2 \right]$$

$$= \frac{W}{YAk^2} \left(\tfrac{1}{3}p^3L^3 + \tfrac{1}{2}p^2L^3 - \tfrac{1}{2}p^3L^3 \right)$$

$$= \frac{W}{YAk^2} \left(\tfrac{1}{2}p^2L^3 - \tfrac{1}{6}p^3L^3 \right)$$

$$= \frac{WL^3}{2YAk^2} \left(p^2 - \tfrac{1}{3}p^3 \right)$$

This proves that the depression at the end of a light cantilever when the load is hung at $x = pL$ is the same as the depression at $x = pL$ when the load is hung at the end.

Heavy cantilever

Consider a heavy beam of length L and of weight S per unit length fixed at one end (Fig. 3.26). The weight to the right of the P acts at the mid-point of the length $L - x$. We have

$$YAk^2y'' = S(L - x).\tfrac{1}{2}(L - x) = \tfrac{1}{2}S(L - x)^2$$
$$= \tfrac{1}{2}S(L^2 + x^2 - 2Lx)$$

Integrating

$$YAk^2y' = \tfrac{1}{2}S(L^2x + \tfrac{1}{3}x^3 - Lx^2) + 0$$

Integrating again

$$YAk^2y = \tfrac{1}{2}S \left(\frac{L^2x^2}{2} + \frac{x^4}{12} - \frac{Lx^3}{3} \right) + 0$$

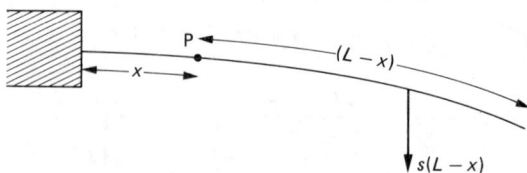

Fig. 3.26

The depression at the end of the beam y_E when $x = L$ is given by

$$YAk^2y_E = \tfrac{1}{2}S \left(\frac{L^4}{2} + \frac{L^4}{12} - \frac{L^4}{3} \right) = \tfrac{1}{8}SL^4$$

This depression is additive to that produced by any hung weights.

Light beam (measurement of Young's modulus)
Consider a light beam loaded at its mid-point by a weight $2W$ and supported at its ends (Fig. 3.27). The reaction at each support is W. This is the problem of the cantilever upside down, and the depression at the centre, or elevation at the ends, $y_E = WL^3/3$. If the beam extends beyond the supports, the elevation at the end may be obtained by using the elevation at the support, together with the slope of the beam at the support.

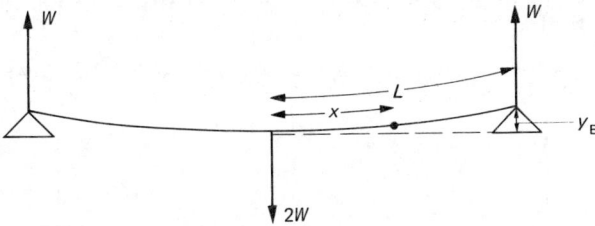

Fig. 3.27

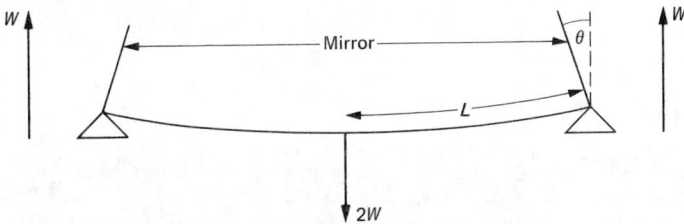

Fig. 3.28

With regard to the measurement of the elastic constants, with the exception of special materials like rubber, the deformations produced are very small, which makes accurate measurement difficult. Bending is normally used to determine Young's modulus Y for wood etc. With reference to Fig. 3.28, mirrors attached to the ends of the beam are tilted from the vertical due to the bending. The slope at the end of the beam y'_E is given by $YAk^2y'_E = \frac{1}{2}WL^2$

and $\quad y'_E = \tan \theta = \theta \quad$ for small angles

θ is measured using an optical method and Young's modulus for the material of the beam is given by

$$Y = \frac{WL^2}{2Ak^2}\frac{1}{\theta}$$

Heavy beam
Consider a heavy beam of length $2L$ and of weight per unit length S supported as shown in Fig. 3.29. SL is the reaction at each support. We have

$$YAk^2y'' = S(L - x).\tfrac{1}{2}(L - x) - SL(L - x)$$

the last term being due to the reaction at the support. Integrating, we find

$$YAk^2y' = \tfrac{1}{2}S(L^2x + \tfrac{1}{3}x^3 - Lx^2) - SL(Lx - \tfrac{1}{2}x^2)$$

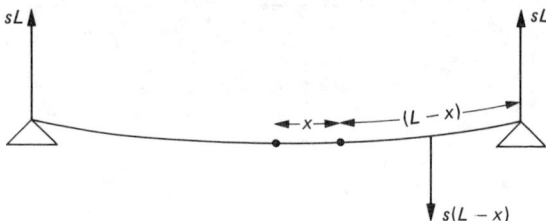

Fig. 3.29

Integrating again:

$$YAk^2y = \tfrac{1}{2}S\left(\frac{L^2x^2}{2} + \frac{x^4}{12} - \frac{Lx^3}{3}\right) - SL\left(\frac{Lx^2}{2} - \frac{x^3}{6}\right)$$

The elevation at the end y_E (or depression at the centre) is given by

$$YAk^2y_E = \tfrac{1}{2}S\left(\frac{L^4}{2} + \frac{L^4}{12} - \frac{L^4}{3}\right) - SL\left(\frac{L^3}{2} - \frac{L^3}{6}\right)$$
$$= \tfrac{1}{8}SL^4 - \tfrac{1}{3}SL^4$$

Light beam (ends loaded two symmetrical supports)
Consider a light beam with a weight W attached to each end and supported as in Fig. 3.30. We have

$$YAk^2y'' = -W(a - x) + W(a + b - x)$$
$$= Wb$$

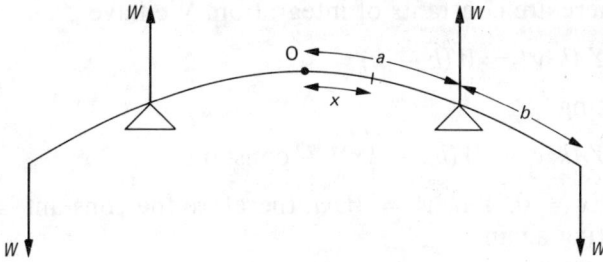

Fig. 3.30

This equation applies for $x < a$, i.e. for the part of the beam between the origin O and the support. It shows that the bending moment is independent of x and a. This means, remembering that $y'' = 1/R$, where R denotes radius of curvature, that the radius of curvature is the same all the way along. A figure with the same radius of curvature everywhere is a circle, therefore the shape of the beam between the supports is the arc of a circle. Integrating

$$YAk^2y' = Wbx + 0$$

For $x = a$ the slope y'_a is given by $YAk^2y'_a = Wba$. Integrating again

$$YAk^2y = Wbx^2/2$$

For $x = a$ the depression y_a is given by

$$YAk^2y_a = Wba^2/2$$

Thus we have found the slope y'_a and the depression of the beam y_a at the support.

Consider now the remaining part of the beam as in Fig. 3.31.

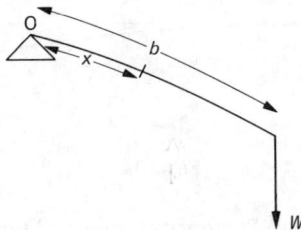

Fig. 3.31

Here there are constants of integration. We have

$$YAk^2y'' = W(b - x)$$

Integrating

$$YAk^2y' = W(bx - \tfrac{1}{2}x^2) + \text{constant}$$

But for $x = 0$, $YAk^2y'_a = Wba$, therefore the constant is Wba. Integrating again

$$YAK^2y = W(\tfrac{1}{2}bx^2 - \tfrac{1}{6}x^3) + Wbax + C$$

where C is a constant. But for $x = 0$

$$YAk^2y_a = Wba^2/2$$

Thus

$$C = Wba^2/2$$

Thus if the depression at the end of the beam is y_b this is given by

$$YAk^2y_b = \frac{Wb^3}{3} + Wb^2a + \frac{Wba^2}{2}$$

Oscillation of light cantilever
Consider a weight W hung at the end of a light cantilever of length L as in Fig. 3.32. The weight W causes a depression at the end y_E given by $YAk^2y_E = WL^3/3$. Thus

$$W = \frac{3YAk^2}{L^3} y_E$$

and the weight is in equilibrium with this particular depression at the end. In general, whenever we have a depression y, opposing it is a force due to the internal resistance of the material, i.e. in general a force $3YAk^2y/L^3$ opposes a depression y.

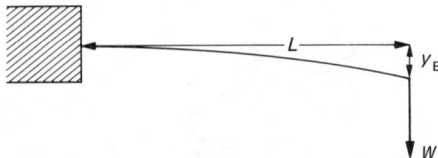

Fig. 3.32

Let $W = Mg$ where M denotes mass. Then if the mass M is pulled down and released the equation of motion of the cantilever is given by

$$M\frac{d^2y}{dt^2} = M\ddot{y} = -\frac{3YAk^2}{L^3}y$$

This equation represents simple harmonic motion and the period T is given by

$$T = 2\pi\sqrt{\frac{ML^3}{3YAk^2}} \quad \text{or} \quad \frac{T^2}{4\pi^2} = \frac{ML^3}{3YAk^2}$$

Measurement of the period of oscillation T thus enables a value for Young's modulus Y to be obtained. If the mass of the beam is not negligible compared with W, a correction must be made to allow for the inertia of the beam.

3.8 Determination of Poisson's ratio and the elastic constants of a wire by Searle's method

The rigidity modulus of a material is best measured with the material in the form of a wire. In Fig. 3.33 two identical brass rods suspended by parallel silk fibres are connected together at their mid-points by a length of the wire under test. The ends of the rods are drawn together slightly by a loop of cotton which is then burnt. A value for the period T_1 of the resulting oscillation is obtained.

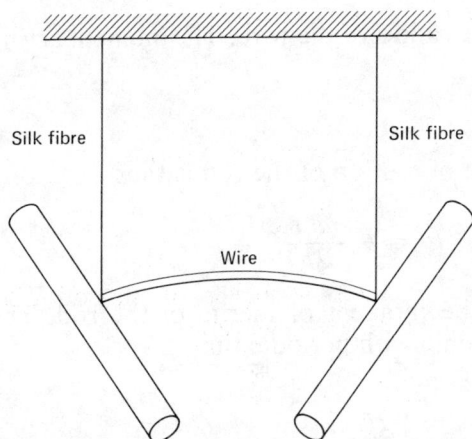

Silk fibre   Silk fibre

Wire

Fig. 3.33

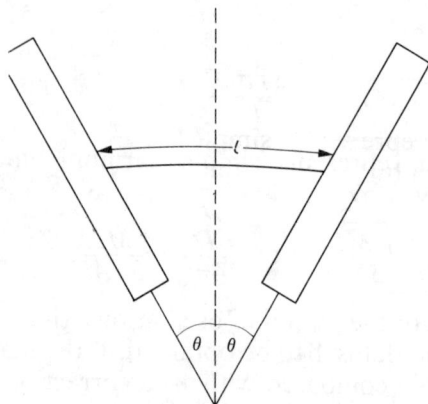

Fig. 3.34

Consider the situation (Fig. 3.34) when each rod is displaced by an angle θ from its equilibrium position. The wire is bent into a circular arc of radius of curvature $l/2\theta$ and the bent wire exerts a restoring torque on each rod equal to the moment of the flexural stresses produced in the wire by the motion of the rod. For a slightly bent wire this moment is YAk^2/R, where Ak^2 is the moment of inertia of the cross section of the wire about the neutral axis, Y is Young's modulus of the wire, and R denotes radius of the curvature. But

$$Ak^2 = \pi a^4/4$$

where a is the radius of the wire. The moment acting is given by

$$\frac{\pi a^4 Y \theta}{2l}$$

The equation of motion of the rod is thus

$$I\frac{d^2\theta}{dt^2} = I\ddot{\theta} = -\frac{\pi a^4 Y}{2l}\theta$$

where I is the moment of inertia of the rod. The motion is simple harmonic with periodic time

$$T_1 = 2\pi\sqrt{\frac{2Il}{\pi a^4 Y}}$$

from which

$$Y = \frac{8\pi I l}{T_1^2 a^4}$$

where

$$I = M\left(\frac{L^2}{12} + \frac{r^2}{4}\right)$$

M is the mass of the rod, L its length and r its radius.

When one rod is clamped in a horizontal position so that the wire can undergo torsional oscillations the periodic time is

$$T_2 = 2\pi\sqrt{\frac{2Il}{\pi a^4 n}}$$

from which the rigidity modulus of the wire is given by

$$n = \frac{8\pi I l}{T_2^2 a^4}$$

Now, from the theory of elasticity, we know that

$$1 + \sigma = \frac{Y}{2n}$$

Therefore

$$\sigma = \left[\frac{1}{2}\left(\frac{T_2^2}{T_1^2}\right) - 1\right]$$

where σ is Poisson's ratio. Having found Y and n, the bulk modulus k of the wire can be calculated from the relation $Y(n + 3k) = 9nk$ or

$$\frac{1}{Y} = \frac{1}{9k} + \frac{1}{3n}$$

3.9 Energy stored in a stretched wire

For the region shown in Fig. 3.35 where Hooke's law holds we know that load is proportional to extension. The work done in increasing the extension by de is given by

average force $\times$ de = $F \times$ de = area of shaded strip

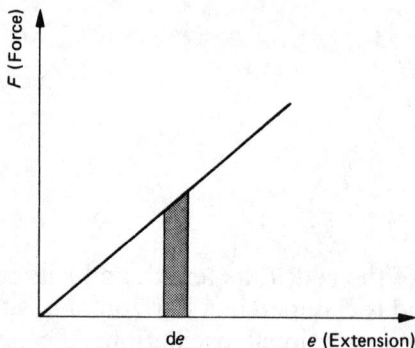

Fig. 3.35

The work done when the extension is e is given by

area under graph $= \frac{1}{2}Fe$

But Young's modulus

$$Y = \frac{F/A}{e/L}$$

where A is the area of cross section of the wire and L its length. Therefore

$$F = YAe/L$$

Thus the energy stored in the wire is

$$\frac{1}{2}YAe^2/L$$

and the energy stored per unit volume is

$$\frac{1}{2}\frac{YAe^2}{L^2A} = \frac{1}{2}Y\frac{e^2}{L^2} = \frac{1}{2}Y(\text{strain})^2$$

But Y is stress/strain and so the energy stored per unit volume is

$$\frac{1}{2}\frac{(\text{stress})^2}{Y} = \frac{1}{2}\text{stress} \times \text{strain}$$

4 Surface Tension

If we consider a liquid as in Fig. 4.1, a molecule in the interior of the liquid is attracted by other molecules, on average, equally in all directions. However, for molecules at the surface the forces are unbalanced, and a molecule in the surface experiences a net force towards the interior of the liquid. This causes a depopulation of the surface, and gives rise to a state of tension in the surface.

Looking at the situation from the point of view of energy, since work is required to extract a molecule from the interior of the liquid, and bring it to the surface, the surface molecules have a high potential energy, and since any system tries to reduce its potential energy, there is a tendency for the surface to contract, i.e. for the surface area to be reduced to a minimum. This can be confirmed experimentally by putting a drop of aniline into water. Aniline has the same density as water (the effect of gravity on the drop is thus eliminated) and it is found that the drop assumes a spherical shape, a sphere being the shape with the smallest surface area for a given volume.

The surface tension S may be defined at a given temperature, as the force per unit length, acting perpendicular to any line element drawn in the surface of the liquid, and acting tangentially to the surface. At constant temperature the work required to separate two parallel lines in the surface, each of length L,

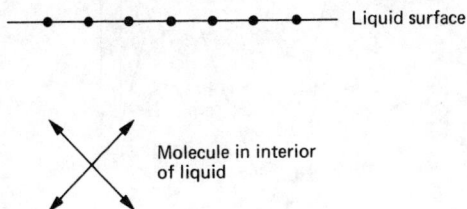

——•—•—•—•—•—•—— Liquid surface

Molecule in interior
of liquid

Fig. 4.1

by dx, is SL dx = SdA, where dA denotes an element of area. Thus a surface is said to contain *free energy* per unit area S. The energy is said to be free as it is available as useful work if the surface is allowed to shrink. However stretched surfaces cool, indicating that in addition some heat energy h per unit area contributes to the total surface energy. Thus the total surface energy per unit area $E = S + h$.

4.1   Pressure difference across a curved surface

ABCD is an element of area (Fig. 4.2) in any curved surface bounding a fluid, whose sides meet at right angles, i.e. ABCD is a curvilinear rectangle. Assume that the surface is in equilibrium with a certain pressure on either side. An excess pressure exists on the concave side to counterbalance the effect of surface tension. Assume that the surface is moved out normally parallel to itself a distance dn:

$$\text{final area} = (\text{AB} + \text{d}n\theta_1)(\text{BC} + \text{d}n\theta_2)$$
$$\text{increase in area} = \text{AB d}n\theta_2 + \text{BC d}n\theta_1 + \text{d}n^2\theta_1\theta_2$$

But

$$\theta_1 = \frac{\text{AB}}{R_1} \quad \text{and} \quad \theta_2 = \frac{\text{BC}}{R_2}$$

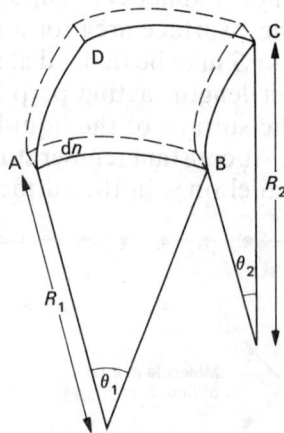

Fig. 4.2

where R_1, R_2 denote the radii of curvature. Thus we have

$$\text{increase in area} = AB.BC\left[\frac{1}{R_1} + \frac{1}{R_2}\right]dn$$

neglecting the small term involving dn^2.

The work done against surface tension S or increase in surface energy is

$$AB.BC.S.dn\left[\frac{1}{R_1} + \frac{1}{R_2}\right]$$

But the work done by the excess pressure P is given by

$$P \times \text{volume increase} = P \times AB.BCdn$$

Therefore

$$P = S\left[\frac{1}{R_1} + \frac{1}{R_2}\right]$$

R_1 and R_2 are the principal radii of curvature of the surface at the point where the element lies. The principal radii of a surface are the ones which are obtained by cutting the surface by means of two planes at right angles, such that one radius is a maximum and the other a minimum.

For a spherical surface, $R_1 = R_2 = R$, where R is the radius of the sphere, and $P = 2S/R$. For a cylindrical surface, $R_1 = R$ and $R_2 = \infty$. Thus $P = S/R$. In the case of a spherical soap bubble, which has two surfaces, $P = 4S/R$, where R is the radius of the bubble.

4.2 Shape of an interfacial boundary under the action of surface tension and gravity

With reference to Fig. 4.3, ρ_1 is the density of the fluid (usually air) on one side of the boundary and ρ_2 is the density of the fluid on the other side of the boundary ($\rho_1 < \rho_2$).

Consider the point A. $P_1 - P_2$ is the pressure difference across the surface at the point A:

$$P_1 - P_2 = S\left(\frac{1}{R_1} + \frac{1}{R_2}\right)_A = C$$

where R_1, R_2 denote radii of curvature at the point A. At the

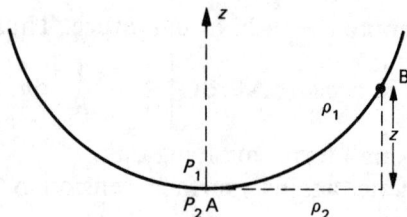

Fig. 4.3

point B, a height z above A, the pressures are $P_1 - \rho_1 gz$ and $P_2 - \rho_2 gz$. Therefore the pressure difference at B is

$$P_1 - P_2 + gz(\rho_2 - \rho_1)$$
$$= C + gz(\rho_2 - \rho_1)$$
$$= S\left(\frac{1}{R_1} + \frac{1}{R_2}\right)_B$$

where R_1, R_2 now denote radii of curvature at the point B. This is a general formula but is difficult to use, so we shall consider some special cases.

4.3   Rise of a liquid against the vertical wall of a glass vessel

We will consider a cylindrical surface (see Fig. 4.4), and ignore the pressure of the atmosphere which, if brought in, cancels out.

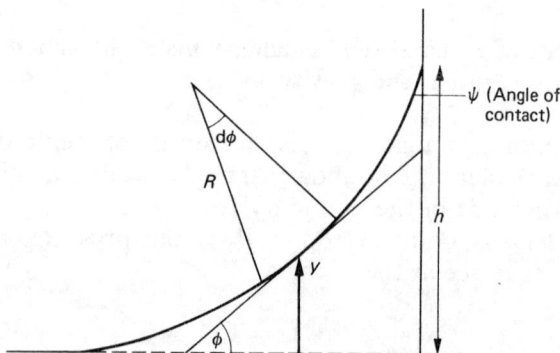

Fig. 4.4

Then the pressure difference is

$$P = \rho g y = S/R$$

Where R denotes the radius of curvature of the surface, and ρ is the density of the liquid of surface tension S. The other radius of curvature is infinite in this case and so the term containing it does not enter in the expression for P.

Now, from Fig. 4.5 we see that $d\phi = ds/R$ where ds denotes an element of the curve. Therefore

$$P = S \frac{d\phi}{ds} = S \frac{d\phi}{dy} \frac{dy}{ds}$$

$$= S \frac{d\phi}{dy} \sin \phi$$

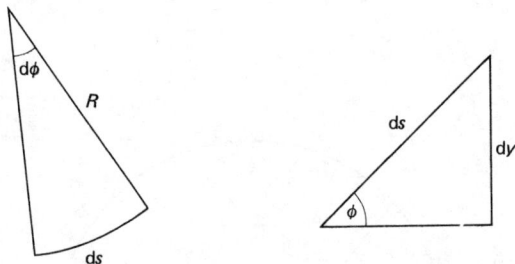

Fig. 4.5

But $P = \rho g y$, therefore

$$\rho g y \, dy = S \sin \phi d\phi$$

which is a separable differential equation. Integrating between the limits $y = 0$ and $y = h$ (Fig. 4.6) we have

$$\int_0^h \rho g y \, dy = \int_0^{90-\psi} S \sin \phi \, d\phi$$

where ψ is the angle of contact with the glass wall. Thus

$$\tfrac{1}{2}\rho g h^2 = -S \big[\cos \phi\big]_0^{90-\psi}$$

$$= -S \big[\sin \psi - 1\big]$$

$$= S \big[1 - \sin \psi\big]$$

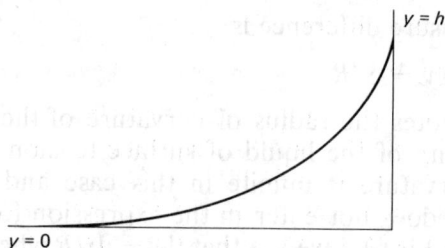

Fig. 4.6

4.4   Sessile drop on horizontal surface

Fig. 4.7 represents the sessile drop; the pressure difference across the surface at the point with coordinate y, as shown, is

$$P = \rho g y = S/R$$

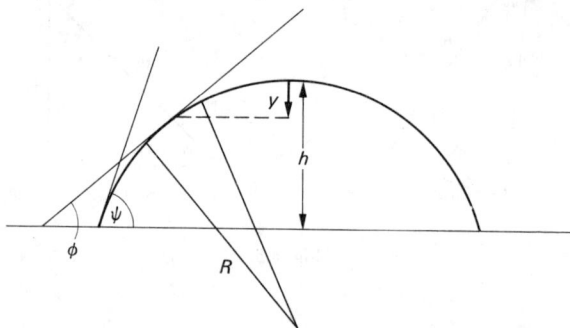

Fig. 4.7

where R denotes radius of curvature, and ρ is the density of the liquid of surface tension S. Again, the other radius of curvature is very much larger than R and so the term in which it occurs can be ignored in our expression for P. From Fig. 4.8 we have $d\phi = ds/R$, where ds is an element of the curve. Therefore

$$P = S \frac{d\phi}{ds} = S \frac{d\phi}{dy} \frac{dy}{ds}$$

$$= S \frac{d\phi}{dy} \sin \phi$$

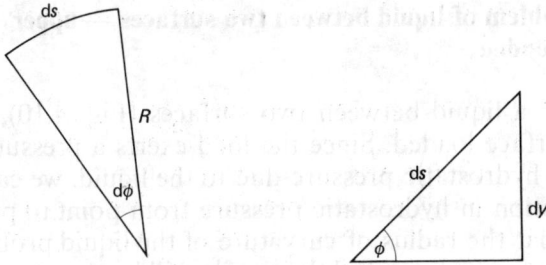

Fig. 4.8

But $P = \rho g y$, therefore

$$\rho g y \, dy = S \sin \phi d\phi$$

and

$$\int_0^h \rho g y \, dy = \int_0^\phi S \sin \phi d\phi$$

where ϕ is the angle of contact with the horizontal surface. Thus

$$\tfrac{1}{2}\rho g h^2 = -S[\cos \phi]_0^\phi$$
$$= -S[\cos \psi - 1] = S(1 - \cos \psi)$$

In the case of the drop shown in Fig. 4.9.

$$\tfrac{1}{2}\rho g h^2 = S$$

since the tangent is vertical and ϕ (Fig. 4.7) is 90°. Measurement of h thus gives a value for the surface tension S. Also, $\tfrac{1}{2}\rho g H^2 = S(1 - \cos \psi)$ and measurement of H gives a value for the angle of contact ψ.

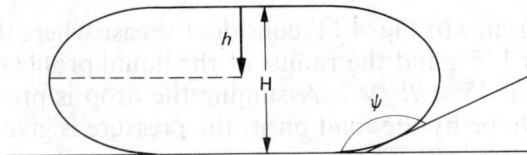

Fig. 4.9

The sessile drop method has been used to measure the surface tension of molten metals over a wide range of temperature.

4.5   Problem of liquid between two surfaces — upper surface loaded

Consider a liquid between two surfaces (Fig. 4.10), with the upper surface loaded. Since the load exerts a pressure greater than the hydrostatic pressure due to the liquid, we can neglect the variation in hydrostatic pressure from point to point. This means that the radius of curvature of the liquid profile will be the same everywhere, and the profile will be the arc of a circle.

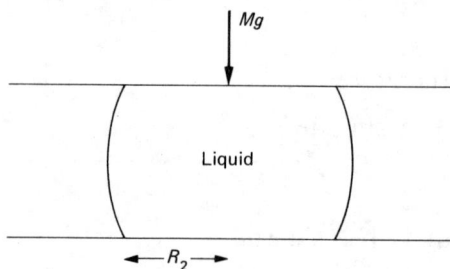

Fig. 4.10

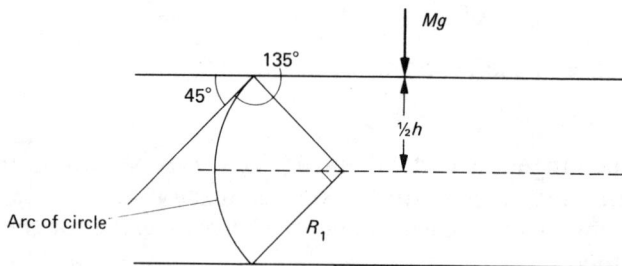

Fig. 4.11

With reference to Fig. 4.11, consider the case where the angle of contact is 135°, and the radius of the liquid profile is R_1. Then $\frac{1}{2}h = R_1 \sin 45 = R_1/\sqrt{2}$. Assuming the drop is pressed into a circular shape by the load on it, the pressure is given approximately by $Mg/\pi R_2^2$. Then

$$\frac{Mg}{\pi R_2^2} = \frac{S}{R_1} = \frac{2S}{h\sqrt{2}} = \frac{S\sqrt{2}}{h}$$

where S is the surface tension of the liquid.

4.6 Rise of a liquid between two parallel vertical plates

In Fig. 4.12, the liquid of density ρ rises between the plates to a height h above the general level. If we assume that the plates are close together and that the angle of contact is zero, the profile curve will be semicircular.

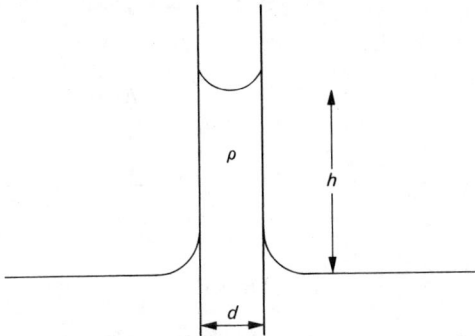

Fig. 4.12

Consider a column of the liquid of unit thickness. The volume of a slice of unit thickness is

$$dh + \frac{d^2}{2} - \frac{\pi d^2}{8}$$

The last two terms represent the area of a rectangle minus the area of a semicircle. The weight of a slice of unit thickness is thus

$$dh\rho g + \rho g d^2 \left(\tfrac{1}{2} - \tfrac{1}{8}\pi\right)$$

But for equilibrium, this weight is equal to the sum of the upward forces due to surface tension S, i.e. to $2S$ (surface tension is defined as force per unit length). Therefore

$$dh\rho g + \rho g d^2 \left(\tfrac{1}{2} - \tfrac{1}{8}\pi\right) = 2S$$

and so

$$h = \frac{2S}{d\rho g} - 0 \cdot 107 d$$

4.7 Measurement of surface tension by capillary rise

The capillary rise method is about the most accurate one for measuring surface tension, and other methods use capillary rise results for calibration purposes.

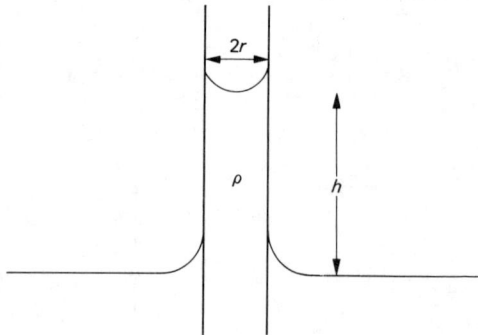

Fig. 4.13

If the capillary tube of radius r (Fig. 4.13) is narrow, and the angle of contact is zero, then we may assume that the meniscus is hemispherical. The upward force due to surface tension S, defined as force per unit length, is $2\pi rS$. But this is equal to the weight of the column of liquid of height h, plus the weight of the meniscus. Therefore

$$2\pi rS = \pi r^2 h\rho g + (\pi r^3 \rho g - \tfrac{2}{3}\pi r^3 \rho g)$$

where ρ is the density of the liquid. The last two terms represent the weight of a cylinder minus the weight of a hemisphere. Thus

$$\frac{2S}{\rho g} = rh + \tfrac{1}{3}r^2$$

$2S/\rho g = a^2$ say, is called the capillary constant and

$$a^2 = rh\left[1 + \tfrac{1}{3}\frac{r}{h}\right]$$

This looks like the first two terms in a power series, and in fact the next terms in the series were worked out by Lord Rayleigh giving

$$a^2 = rh\left[1 + \tfrac{1}{3}\frac{r}{h} + A\left(\frac{r}{h}\right)^2 + B\left(\frac{r}{h}\right)^3 + \dots\right]$$

($r \ll h$, i.e. for very narrow tubes) where Rayleigh's values for A and B were $A = -0 \cdot 1288$ and $b = +0 \cdot 1312$ for $r/h < 0 \cdot 25$. Thus the determination of the surface tension of a liquid by capillary rise requires measurements of r and of h.

Points to be considered for an accurate determination of surface tension by capillary rise
Firstly, the angle of contact must be zero otherwise the method is not a precision one. An optical method may be used to ensure that the liquid has zero angle of contact.

Secondly, h is measured by means of a travelling microscope of high precision, with careful illumination of the meniscus, which is important. Some sort of marker, like a pin point, is needed to locate the surface.

Thirdly, the capillary tube must not be placed in another container which is too narrow.

4.8 Measurement of surface tension by Ferguson's method

This method is used when only a small quantity of liquid is available and its density cannot be determined.

A small quantity of the liquid under investigation is placed in a clean horizontal capillary (Fig. 4.14), which is connected to a sensitive manometer. The pressure inside the apparatus can be gradually increased, by means of a syphon bottle supported on a rising table.

Light from a clear filament lamp is focussed on the exposed end of the liquid column, which may be convex, plane, or concave according to the pressure within the apparatus. The pressure inside the apparatus is adjusted until the end is plane, which is the case when the image of the lamp as seen in the microscope, opens out to a pool of light across the liquid surface.

The surface tension force $2\pi rS$ (assuming zero angle of contact) at the other curved meniscus, must then equal $\rho gh\pi r^2$, where r is the radius of the capillary.

Thus $2S/r = \rho gh$, where ρ is the density of the manometric fluid, and the surface tension of the liquid under investigation S, can be determined. The experiment can then be repeated using a number of capillaries of different radii.

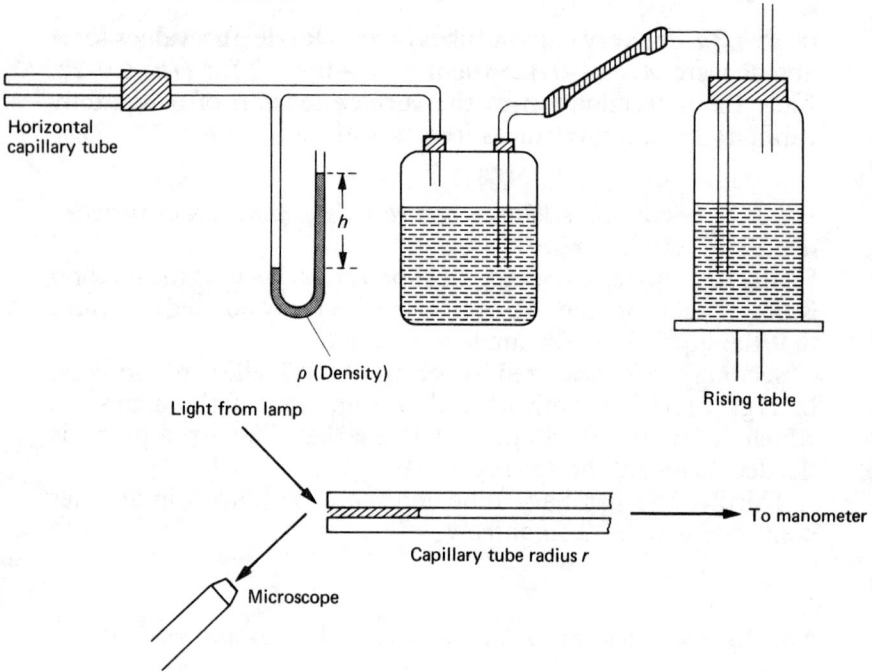

Fig. 4.14

4.9   Measurement of surface tension by the drop weight method

The liquid whose surface tension is required, is allowed to form drops at the end of a capillary tube of outer radius r (Fig. 4.15). Consider the equilibrium of the drop when it is just about to fall:

$$\text{downward force} = Mg + \frac{S}{r}\pi r^2$$

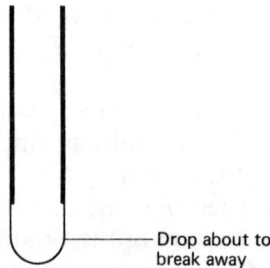

Fig. 4.15

where Mg is the weight of the drop, and S denotes surface tension. The excess pressure inside the drop is $S[(1/r_1)+(1/r_2)]$, but r_2 is infinite since the drop is of cylindrical cross section. Hence the excess pressure is $S/r_1 = S/r$:

upward force $= 2\pi r S$

Hence for equilibrium

$$2\pi r S = Mg + \frac{S}{r}\pi r^2 \quad \text{or} \quad S(2\pi r - \pi r) = Mg$$

Therefore $Mg = \pi r S$. Thus for a given capillary $Mg = KS$, K being a constant. However, this is deduced assuming static conditions which do not hold during the experiment. Also, the assumption of cylindrical shape is not satisfactory.

Harkins and Brown have shown that K is not constant, even for a given capillary tube, but that $Mg = rSf$, where f is a function of $r/V^{1/3}$, where V is the volume of the drop. That is,

$$f = 2\pi\phi\left(\frac{r}{V^{1/3}}\right) \quad \text{and} \quad Mg = 2\pi r S\phi$$

The drop weight is obtained by collecting about thirty drops in a specific gravity bottle. From M and the density of the liquid,

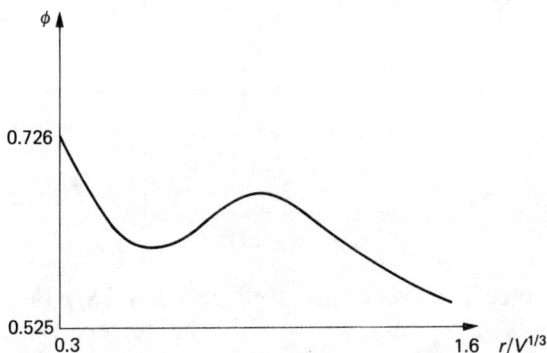

Fig. 4.16

$V^{1/3}$ and hence $r/V^{1/3}$ is calculated. The value of $\phi(r/V^{1/3})$ is taken from the graph of Harkins and Brown shown in Fig. 4.16, and S is calculated from $Mg = 2\pi r S\phi$. Success in the accurate use of the method depends amongst other things on a slow drop formation, a perfectly circular end to the tube, and the use of a thick-walled tube.

4.10   Surface tension and vapour pressure

The maximum steady value of the pressure of vapour molecules above a liquid surface, is the saturated vapour pressure or vapour pressure. The number of molecules leaving the surface per second is then equal to the number returning to it per second. The vapour pressure depends in addition to temperature, on the shape of the surface. It is greater over a convex surface than over a plane surface, since a vapour molecule can escape more readily from a convex surface, as the effect of the attraction of neighbouring liquid molecules is less. Conversely the vapour pressure over a concave surface is less than over a plane one.

To obtain the relationship between surface tension and vapour pressure, consider a liquid in contact with its vapour in a completely closed vessel (Fig. 4.17). A capillary tube of radius r is placed upright in the liquid. P is the saturated vapour pressure at the plane surface, and $\bar{P}$ is that over the convex surface. σ is the density of the vapour and ρ that of the liquid. Now

$$P - \bar{P} = \sigma g h$$

But $P + (2S/r)$ is the pressure just inside the curved liquid

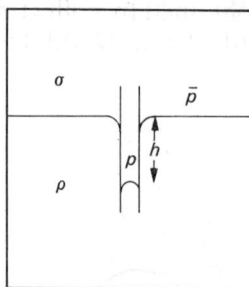

Fig. 4.17

surface, since the excess pressure inside is $2S/r$ (assuming the meniscus is spherical), where S is the surface tension of the liquid. However, this must equal the pressure at the same level outside the capillary, i.e.

$$P + \frac{2S}{r} = \bar{P} + \rho g h$$

Therefore $P - \bar{P} = \sigma g h = \rho g h - \dfrac{2S}{r}$

and so $\dfrac{2S}{r} = (\rho - \sigma)gh = \dfrac{\rho - \sigma}{\sigma}(P - \bar{P})$

since $P - \overline{P} = \sigma g h$. Thus the vapour pressure over a convex surface of radius r is greater than that over a plane surface, by an amount

$$\frac{2S}{r} \frac{\sigma}{\rho - \sigma}$$

This assumes that h is small enough to make the variation in vapour density with height negligible.

We have assumed that the meniscus is spherical, and we now use a second approximation, i.e. that the perfect gas equation can be used for a vapour.

In this case

$$P = \sigma \frac{RT}{M}$$

where R is the universal gas constant, T the absolute temperature, and M the molecular weight of the vapour. Therefore

$$\frac{2S}{r} = \int_{\overline{P}}^{P} \frac{\rho - \sigma}{\sigma} dP \cong \int_{\overline{P}}^{P} \frac{\rho}{\sigma} dP$$

since $\sigma \ll \rho$. But $\dfrac{1}{\sigma} = \dfrac{RT}{M} \dfrac{1}{P}$

Therefore

$$\frac{2S}{r} = \rho \frac{RT}{M} \int_{\overline{P}}^{P} \frac{dP}{P} = \rho \frac{RT}{M} \log \left(\frac{P}{\overline{P}} \right)$$

or

$$\frac{2S}{\rho r} = \frac{RT}{M} \log \left(\frac{P}{\overline{P}} \right)$$

Applying this result to liquid droplets, the smaller the drop, the larger the equilibrium vapour pressure P required to counter the tendency to evaporate. For a given value of P, r is the drop radius for equilibrium, and a smaller value of r will lead to the disappearance of the drop by evaporation. Conversely a larger value of r will mean that the drop will grow in size. Thus for a given value of P, if drops are to form at all, they must have a radius of at least that given by the above equation. Thus small particles of matter present in a supersaturated vapour will assist condensation, since the degree of curvature required is less, if such a particle acts as a nucleus for condensation.

5 Viscosity

Viscosity is the property of a fluid whereby it offers a resistance to the passage of a solid body through it. The resistance is due to the impeding effect of molecular collisions. In the case of some liquids, e.g. treacle and glycerin, the resistance is considerable, and such liquids are said to be viscous. Gases also show the effect but are much less viscous.

Turbulent flow   Streamline flow

Fig. 5.1

In the theory we shall deal with, we make the preliminary assumption that the fluid flow is 'streamline', i.e. that the molecules travel along parallel lines (not necessarily straight lines). For non-streamline or 'turbulent' flow, eddy currents are set up (Fig. 5.1), and this occurs when the average flow velocity exceeds a certain critical value.

5.1 Newton's law of viscous flow (non-turbulent motion)

Whenever we have a line of liquid flow with adjacent lines of flow having different velocities (Fig. 5.2), there is a tangential frictional force retarding that flow. Newton's law of viscous flow

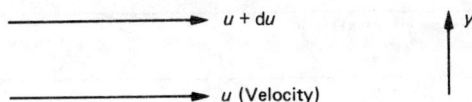

Fig. 5.2

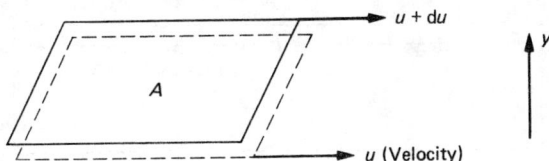

Fig. 5.3

states that the frictional force is proportional to the rubbing area multiplied by the velocity gradient perpendicular to the direction of flow. The rubbing area A is the area over which two adjacent liquid surfaces move with relative velocity (Fig. 5.3); it is not the cross-sectional area. Thus

$$\text{frictional force} = \eta \times \text{rubbing area} \times \frac{du}{dy}$$

where η is called the coefficient of viscosity of the fluid. This law holds only for streamline flow. The CGS unit for the coefficient of viscosity η is the poise (g cm^{-1}s^{-1}). In the SI system the unit of η is kg m^{-1} s^{-1}, hence 1 poise = $0 \cdot 1$ kg m^{-1} s^{-1}. η varies enormously from a very low value, $0 \cdot 18 \times 10^{-4}$ kg m^{-1} s^{-1} for air, to 140 kg m^{-1} s^{-1} for treacle, or approximately 10^9 kg m^{-1} s^{-1} for pitch. Other values for viscosity are 1×10^{-3} kg m^{-1}s^{-1} for water, and $0 \cdot 85$ kg m^{-1} s^{-1} for glycerin. All these values are at room temperature.

5.2 The flow of an incompressible liquid through a narrow tube (Poiseuille's equation)

Assume that the liquid flows fairly slowly through a horizontal tube of internal radius a, and of length L (Fig. 5.4), the flow being maintained by a pressure difference P, applied externally between the ends of the tube.

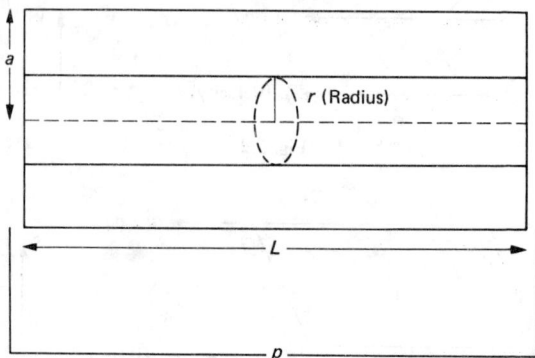

Fig. 5.4

We shall make a number of assumptions:
(1) the flow is streamline (no eddies);
(2) the fluid is at rest at the walls of the tube;
(3) steady-state conditions exist, i.e. there is no acceleration
of the fluid.

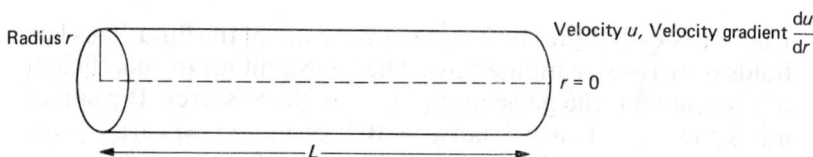

Fig. 5.5

Consider a cylinder of the fluid of radius r (Figures 5.4 and 5.5).
A force $\pi r^2 P$ drives this cylinder past a rubbing area $2\pi rL$. Thus,
equating the driving force to the viscous drag, we have

$$\pi r^2 P = \eta (2\pi rL)\left(-\frac{du}{dr}\right)$$

by Newton's law, where du/dr is the velocity gradient at a
distance r from the axis of the tube, the velocity is u, and η
is the viscosity of the fluid. The negative sign arises since, as
r increases the velocity u decreases, making du/dr negative

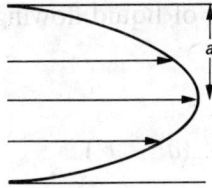

Velocity profile for a liquid flowing
through a narrow tube

Fig. 5.6

(Fig. 5.6). Therefore

$$Pr = -\eta(2L)\frac{du}{dr}$$

and

$$Pr\, dr = -2\eta L\, du$$

Integrating, we have

$$P\int_r^a r\, dr = -2\eta L \int_u^0 du$$

The upper limits of the integrals arise because the velocity is
zero at the walls of the tube. Therefore

$$P(a^2 - r^2)/2 = -2\eta L(0 - u)$$
$$= 2\eta Lu$$

Therefore the velocity at a distance r from the axis of the tube is

$$u = \frac{P}{4\eta L}(a^2 - r^2)$$

which indicates a parabolic velocity profile across the tube
(see Fig. 5.6). Now consider the flow of fluid, which has velocity
u, in a sleeve (volume contained between cylinders radius r and
$r + dr$) of radius r (Fig. 5.7). The volume $2\pi r\, dr \times u$ moves past
any given plane in one second.

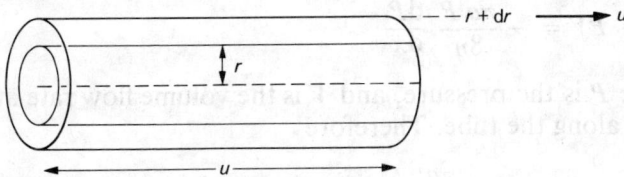

Fig. 5.7

Therefore the volume of liquid flowing per second through the sleeve is

$$\mathrm{d}V = 2\pi r\, \mathrm{d}r\, u$$

$$= 2\pi r\, \mathrm{d}r\, \frac{P}{4\eta L}\, (a^2 - r^2)$$

$$= \frac{\pi P}{2\eta L}\, (a^2 r - r^3)\mathrm{d}r$$

Integrating to get the total volume of liquid V flowing through the tube per second, we find

$$V = \int \mathrm{d}V = \int_0^a \frac{\pi P}{2\eta L}\, (a^2 r - r^3)\mathrm{d}r$$

$$= \frac{\pi P}{2\eta L}\, (\tfrac{1}{2}a^4 - \tfrac{1}{4}a^4) = \frac{\pi P a^4}{8\eta L}$$

5.3   Modification of Poiseuille's equation for gases

Poiseuille's equation holds for an incompressible fluid, but in the case of a gas the pressure gradient $\mathrm{d}P/\mathrm{d}x$ and the volume flow rate V are separately functions of x (x is distance along the tube).

Thus we write Poiseuille's equation for an element of length $\mathrm{d}x$:

$$V(x) = -\frac{\pi a^4}{8\eta}\frac{\mathrm{d}P}{\mathrm{d}x}$$

The negative sign is introduced since, as x increases, the pressure P decreases, thus making $\mathrm{d}P/\mathrm{d}x$ negative.

However, the mass flow rate is constant along the tube (streamline flow), and this is proportional to PV (Boyle's law), i.e. PV is not a function of x but is constant. Thus

$$PV = -\frac{\pi a^4 P}{8\eta}\frac{\mathrm{d}P}{\mathrm{d}x}$$

where P is the pressure, and V is the volume flow rate at some point along the tube. Therefore

$$PV\, \mathrm{d}x = -\frac{\pi a^4}{8\eta} P\, \mathrm{d}P$$

Integrating, we have

$$\int_0^L PV \, dx = -\frac{\pi a^4}{8\eta} \int_{P_1}^{P_2} P \, dP$$

Where L is the length of the tube, P_1 the pressure at one end and P_2 the pressure at the other. Evaluating these integrals, we get

$$PVL = -\frac{\pi a^4}{8\eta} \left[\frac{P^2}{2} \right]_{P_1}^{P_2}$$

or

$$PV = \frac{\pi a^4}{16\eta L} (P_1^2 - P_2^2) = P_1 V_1 = P_2 V_2$$

where V_1 is the volume flow rate at one end of the tube, and V_2 is that at the other. Since

$$P_1^2 - P_2^2 = (P_1 - P_2)(P_1 + P_2) \cong 2P(P_1 - P_2)$$

if the gas is not very compressible. Thus, in this case the unmodified Poiseuille's equation is obtained.

5.4 Problem of flow through a tube when the pressure head diminishes with time

Most problems on Poiseuille's equation involve a manipulation of the pressure which drives the fluid.

With reference to Fig. 5.8, the driving pressure is derived from a head of fluid which is not constant, but which diminishes

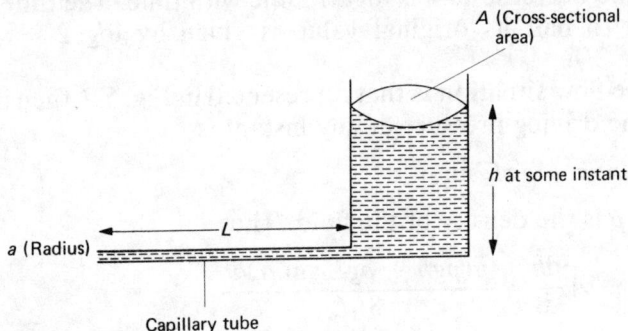

Fig. 5.8

with time. At any instant the driving pressure equals the hydro-static pressure ρgh, where ρ is the density of the fluid. The volume rate of flow V, equals the rate of loss of liquid from the tank, i.e.

$$V = -A \frac{dh}{dt}$$

the negative sign arises since h decreases with time t. Thus

$$-A \frac{dh}{dt} = \frac{\pi \rho gha^4}{8\eta L}$$

where η is the viscosity of the fluid. This is a separable differential equation and we have

$$\frac{dh}{h} = -\frac{\pi \rho ga^4}{8\eta LA} dt$$

If we let $\pi \rho ga^4/8\eta LA = \mu$, we have

$$\frac{dh}{h} = -\mu \, dt$$

Integrating:

$$\int_{h_1}^{h_2} \frac{dh}{h} = -\int_0^t \mu \, dt$$

where h_1 is the height of the fluid at $t = 0$, and h_2 that at time t, i.e.

$$\log \left(\frac{h_2}{h_1}\right) = -\mu t \quad \text{or} \quad \log \left(\frac{h_1}{h_2}\right) = \mu t$$

Thus the decrease in h is logarithmic with time. The time for h to fall to half its original value is given by $\log 2 = \mu t$ or $t = \log 2/\mu$.

If the flow situation is that represented in Fig. 5.9, then in this case the driving pressure at any instant is

$$\rho gh + \rho gL \sin \alpha$$

where ρ is the density of the fluid. Thus

$$-A \frac{dh}{dt} = \frac{\pi(\rho gh + \rho gL \sin \alpha)a^4}{8\eta L}$$

where A is the area of cross section of the tank. From this we

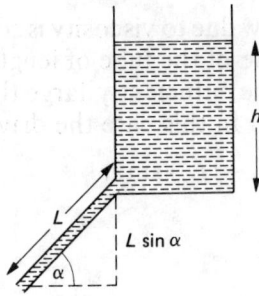

Fig. 5.9

have

$$\frac{-\mathrm{d}h}{\mathrm{d}t} = \frac{\pi\rho g(h + L\sin\alpha)a^4}{8\eta LA}$$

$$= \mu(h + L\sin\alpha)$$

Therefore

$$\frac{\mathrm{d}h}{h + L\sin\alpha} = -\mu\mathrm{d}t$$

Integrating, we obtain

$$\log\left(\frac{h_1 + L\sin\alpha}{h_2 + L\sin\alpha}\right) = \mu t$$

since $L\sin\alpha$ is a constant, where h_1 is the height of the fluid at $t = 0$, and h_2 that at time t.

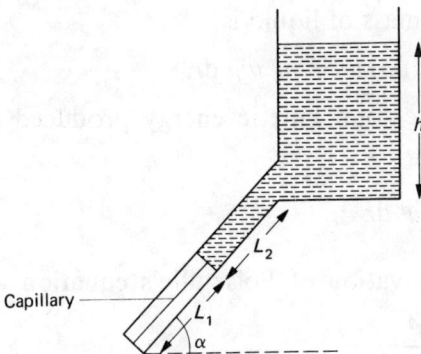

Capillary

Fig. 5.10

The resistance to flow due to viscosity is sensitive to the radius of the tube. In the case of the tube of length L_2 (Fig. 5.10), the radius is assumed to be sufficiently large that there is virtually no viscosity resistance. In this case the driving pressure

$$P = \rho g[h + (L_1 + L_2) \sin \alpha]$$

and

$$- A \frac{dh}{dt} = \frac{\pi P a^4}{8 \eta L_1}$$

5.5   Corrections to Poiseuille's equation

For accurate work two corrections, which have so far been ignored, are required to Poiseuille's equation. First, non-uniform flow occurs at the ends of the tube. This can be corrected for by replacing L in the equation, by $L + Ka$, where K is to be approximately 1.64, and a is the radius of the tube. Secondly, the pressure difference P between the ends of the tube, is not only used to overcome viscous resistance, but is also partly used in imparting kinetic energy to the fluid.

To calculate the kinetic energy produced per second, consider the sleeve of volume equal to that contained between cylinders, of radius r and $r + dr$, as in Fig. 5.7. The volume of liquid flowing through this sleeve per second is $2\pi r \, u \, dr$, where u is the velocity of the liquid at a distance r from the axis of the tube, and the mass of this volume is $2\pi r u \rho \, dr$, where ρ is the density of the liquid. Thus the kinetic energy produced per second in this mass of liquid is

$$\tfrac{1}{2}(2\pi r \, dr \, u\rho) \, u^2 = \pi r \, u^3 \rho \, dr.$$

Integrating, the total kinetic energy produced per second is given by

$$\pi \rho \int_0^a r \, u^3 \, dr$$

But in the derivation of Poiseuille's equation we found that

$$u = \frac{P(a^2 - r^2)}{4 \eta L}$$

Therefore the kinetic energy produced per second is

$$\frac{\pi \rho P^3}{(4\eta L)^3} \int_0^a (a^2 - r^2)^3 \, r \, dr$$

$$= \frac{\pi \rho P^3}{(4\eta L)^3} \int_0^a (a^6 r - 3a^4 r^3 + 3a^2 r^5 - r^7) \, dr$$

$$= \frac{\pi \rho P^3}{(4\eta L)^3} \left[\frac{a^8}{2} - \frac{3a^8}{4} + \frac{3a^8}{6} - \frac{a^8}{8} \right]$$

$$= \frac{\pi \rho P^3}{(4\eta L)^3} \frac{a^8}{8}$$

But the volume flow rate $\quad V = \dfrac{\pi P a^4}{8\eta L}$

Therefore $\quad \dfrac{P}{4\eta L} = \dfrac{2V}{\pi a^4}$

and so $\quad \left(\dfrac{P}{4\eta L} \right)^3 = \dfrac{8V^3}{\pi^3 a^{12}}$

Therefore, the kinetic energy produced per second is

$$\pi \rho \, \frac{8V^3}{\pi^3 a^{12}} \frac{a^8}{8} = \frac{\rho V^3}{\pi^2 a^4}$$

Now the measured pressure difference P, may be written $P = P_1 + P_2$, where P_1 is the part of P required to overcome forces due to viscosity, and P_2 is the part required to produce kinetic energy. The work done by P_2 per second is $P_2 \pi a^2 \times$ average velocity of liquid. But the average velocity of the liquid is $V/\pi a^2$. Therefore the work done by P_2 per second is $P_2 V$, which is equal to the kinetic energy produced in the liquid per second, i.e.

$$P_2 V = \frac{\rho V^3}{\pi^2 a^4} \quad \text{and} \quad P_2 = \frac{\rho V^2}{\pi^2 a^4}$$

Thus when using Poiseuille's equation the measured pressure difference P, should be replaced by

$$P_1 = P - \frac{\rho V^2}{\pi^2 a^4}$$

However, for long tubes of fine bore use of the uncorrected pressure P, gives results of fair accuracy.

5.6   Measurement of the viscosity of a liquid by capillary flow

With reference to Fig. 5.11, the volume of liquid V flowing through the capillary tube per second is determined by weighing the liquid collected in a known time. The pressure difference P between the ends of the tube is ρgh (constant because of the constant-head device), where ρ is the density of the liquid. Thus using Poiseuille's equation, η, the coefficient of viscosity of the liquid, is $\pi Pa^4/8LV$, where a is the radius of the capillary and L its length.

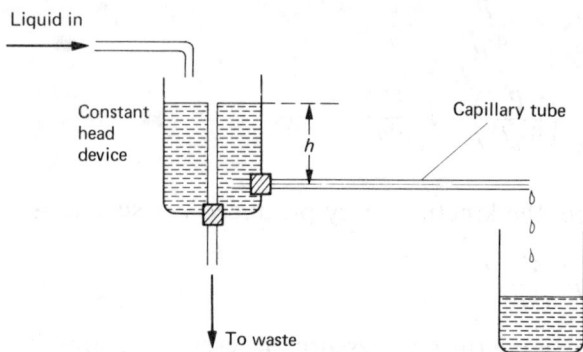

Fig. 5.11

Since the fourth power of the small quantity a is used to obtain η, a must be determined with some accuracy. This may be done by introducing a column of mercury into the capillary, and measuring its length l_1, with a travelling microscope (Fig. 5.12). The majority of the mercury is then poured out and weighed (mass M). The length l_2 of the small bead remaining is then measured and subtracted from the original length, to find the length l of the mercury column that was weighed. Clearly

$$\rho_{Hg} = \frac{M}{\pi a^2 l}$$

and

$$a^4 = \left(\frac{M}{\pi l \rho_{Hg}}\right)^2$$

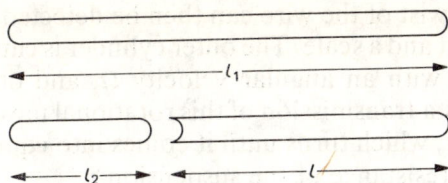

Fig. 5.12

In this way end corrections are avoided.

If necessary, to take account of the kinetic energy of the issuing liquid, the correction $\rho V^2/\pi^2 a^4$ may be applied to P. Finally, in order to ensure streamline flow and no turbulence, the rate of flow of the liquid should not be too great. Check that $P \propto V$.

5.7 The rotation viscometer

The rotation viscometer has various forms, but in its simplest form (Fig. 5.13), an outer cylinder of radius a, containing the liquid, rotates. Concentric with it is a solid inner cylinder of radius b, suspended by means of a wire, attached to which is a

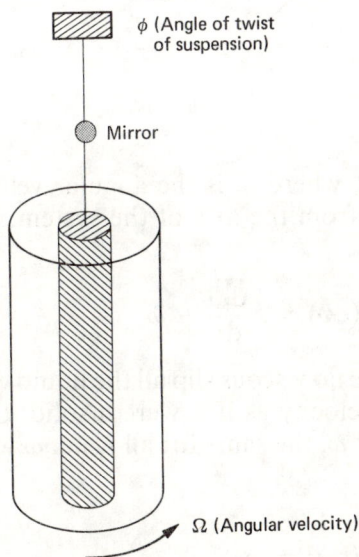

Fig. 5.13

mirror. Any twist of the wire can then be detected by means of a beam of light and a scale. The outer cylinder is caused to rotate mechanically with an angular velocity Ω, and due to viscous forces there is a transmission of this rotational movement to the inner cylinder, which turns until it comes into equilibrium with the torsional resistance of the suspension.

We shall consider the steady state, where there is a definite and constant angle of twist of the torsion fibre, and we shall consider the liquid at a distance r from the axis of the system (Fig. 5.14). This liquid undergoes circular motion with a tangential velocity v, which increases as r increases. For $r = b$ the liquid is at rest, while for $r = a$ the liquid has an angular velocity Ω, i.e. a tangential velocity $a\Omega$.

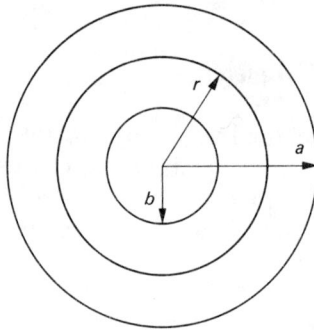

Fig. 5.14

Now $v = r\omega$, where ω is the angular velocity of the liquid at a distance r from the axis of the system. Thus the velocity gradient

$$\frac{dv}{dr} = \frac{d}{dr}(r\omega) = r\frac{d\omega}{dr} + \omega$$

But if there were no viscous slip all the liquid would rotate at the same angular velocity as if it were a solid; there would be an angular velocity ω, the same for all r, i.e. ω would be constant, and

$$\frac{dv}{dr} = \frac{d}{dr}(r\omega) = \omega$$

Thus for the viscous case where

$$\frac{dv}{dr} = r\frac{d\omega}{dr} + \omega$$

the second term ω represents the purely geometrical velocity gradient of circular motion when there is no slipping, and r $d\omega/dr$ is the velocity gradient due to viscous effects which must be used in Newton's law:

viscous force $= \eta \times$ rubbing area $\times$ velocity gradient

Thus for the layer of liquid at the surface of a cylinder of radius r, the circumferential force due to viscosity is

$$\eta(2\pi r)Lr\frac{d\omega}{dr}$$

where η is the coefficient of viscosity of the liquid, and L the length of the inner cylinder covered by liquid. The torque, G say, is given by the force acting multiplied by r. Thus

$$G = \eta(2\pi r)Lr^2\frac{d\omega}{dr}$$

i.e.

$$G = 2\pi\eta Lr^3\frac{d\omega}{dr}$$

In the steady state, when the velocity of each part of the liquid has become constant, this torque G is a constant independent of r, and is equal to the opposing torsional torque $C\phi$, produced by the suspension wire, where C is the torsional constant of the suspension and ϕ its angle of twist.

Integrating, we have

$$G \int_b^a \frac{dr}{r^3} = 2\pi\eta L \int_0^\Omega d\omega$$

Therefore

$$-G\left[\frac{1}{2r^2}\right]_b^a = 2\pi\eta L\Omega$$

or

$$G\left[\frac{1}{b^2} - \frac{1}{a^2}\right] = 4\pi\eta L\Omega$$

The coefficient of viscosity η is obtained by arranging that the outer cylinder goes round slowly enough to avoid turbulence but fast enough to give a measurable angle of twist ϕ to the suspension. Measurement of ϕ gives G since $G = C\phi$ and hence η can be determined. In practice C is not obtained from elasticity theory but from the time of free oscillation of the inner cylinder. The period of oscillation $T = 2\pi\sqrt{I/C}$, where I is the moment of inertia of the inner cylinder about the axis of suspension.

The rotation viscometer (practical details)
The method of measuring η using a rotation viscometer is available for quite viscous liquids, e.g. oil, treacle, and for less viscous liquids like water. It can even be used for gases. However, for gases it is better to use a flow tube method.

The gap between the inner and outer cylinders must be small (one or two mm), since the torque on the inner cylinder is not very great if there is a lot of liquid between the cylinders. High-speed rotation of the outer cylinder is required for gases, or less viscous liquids.

In the case of viscous liquids, high-speed rotation gives a large angle of twist ϕ, but also gives turbulence. Hence the speed of rotation must be as high as possible without running the risk of turbulence.

The formula was derived assuming no awkward and effects, but in fact a viscous torque acts over the base of the inner cylinder. For accurate work account must be taken of end effects. If these are independent of L, they may be eliminated by using two different lengths of inner cylinder immersed.

Elimination of end effects (rotation viscometer)
End effects are eliminated by arranging for the same values of Ω, but different values of ϕ, for different values of L. Now $\phi \propto L$ without any end effects. Taking into account end effects, we have

$$\phi_1 = kL_1 + E$$

where E is the error and

$$k = \frac{4\pi\eta\Omega}{C[(1/b^2) - (1/a^2)]}$$

Also

$$\phi_2 = kL_2 + E$$

assuming the error is the same for a different length. Therefore

$$\phi_1 - \phi_2 = k(L_1 - L_2)$$

Hence k is determined, and knowledge of k gives the viscosity η.

5.8 The falling-sphere viscometer

Consider a sphere dropped into a liquid of coefficient of viscosity η. Initially the sphere will accelerate under gravity, but as its velocity increases so does the viscous resistance offered by the liquid. Eventually the viscous resistance will be equal and opposite to the effective weight of the sphere, and it will then travel at a constant velocity, its 'terminal velocity'.

For a sphere of radius a travelling with terminal velocity v through a fluid of infinite extent, the viscous retarding force F was derived by Stokes as $F = 6\pi\eta a v$. Now the net downward force acting on the sphere is its weight $\frac{4}{3}\pi a^3\rho g$, where ρ is its density, minus the upthrust on it due to Archimedes' principle, $\frac{4}{3}\pi a^3 \sigma g$ (weight of liquid displaced), where σ is the density of the liquid. Thus when the sphere has acquired its terminal velocity,

$$6\pi\eta a v = \tfrac{4}{3}\pi a^3 (\rho - \sigma)g$$

The coefficient of viscosity η may be obtained using this equation, by measuring the time of fall of a steel sphere between two marks on a tube containing the liquid. The upper mark must be sufficiently far down the tube to ensure that the sphere has acquired its terminal velocity before reaching the mark. Once a value for v has been obtained, measurement of a, ρ and σ enables η to be determined. Since η is very sensitive to temperature, the whole apparatus should be enclosed in a constant-temperature bath to ensure a uniform temperature (temperature should always be quoted when a measurement of η is made).

Now Stokes' law was derived for an infinite fluid, thus a correction must be applied when using a fall tube of radius R. This correction has been given as $v_\infty = v[1 + 2\cdot4(a/R)]$ by Ladenburg, where v is the observed terminal velocity, and v_∞

is the terminal velocity for an infinite fluid. A less important correction to allow for the finite length of the tube may also be applied, i.e. $v_\infty = v[1 + 3 \cdot 3(a/h)]$, where h is the height of the liquid in the fall tube.

Certain advances have been made in the theory of viscosity by using the method of dimensions. This will be dealt with in the next chapter.

6 Dimensional Analysis

We will restrict our discussion only to the units which arise in mechanics. A physical quantity is made up of two parts, a number and a unit. The number results from an operation of comparison, and the unit tells us what the comparison has been made against. In mechanics the fundamental magnitudes are mass M, length L, and time T, and due to the progression of science more complicated magnitudes have been introduced. These derived magnitudes are products or divisions of the fundamental magnitudes, e.g.

$$[\text{velocity}] = \frac{[\text{length}]}{[\text{time}]} = LT^{-1}$$

$$[\text{acceleration}] = \frac{[\text{length}]}{[\text{time} \times \text{time}]} = LT^{-2}$$

where the square brackets denote the dimensions of the quantity they contain.

Force = mass × acceleration involves four operations, i.e.

$$[\text{force}] = MLT^{-2}$$

and energy or work is

$$[\text{energy}] = ML^2T^{-2}$$

In mechanics all the derived magnitudes are combinations of the fundamental magnitudes based on the same units, e.g. MKS. In heat and electricity other units have been introduced, but with these we shall not be concerned.

Dimensionless ratios result from comparisons of similar magnitudes and are pure numbers. They are not part of the theory. Sums and differences (necessarily of similar magnitudes) require one operation only in principle, and count as one magnitude. Some operations contain redundancies which may be removed from the dimensional formula, e.g.

$$[\text{pressure}] = \frac{[\text{force}]}{[\text{area}]}$$

$$= \frac{MLT^{-2}}{L^2} = \frac{MLL^{-1}T^{-2}}{L} = ML^{-1}T^{-2}$$

$$[\text{viscosity}] = \frac{[\text{force}]}{[\text{area}][\text{velocity gradient}]}$$

$$= \frac{MLT^{-2}}{L^2 \times LT^{-1}/L} = ML^{-1}T^{-1}$$

Thus in general we simply use the power laws of algebra in dealing with the exponents of M, L and T.

6.1   Uses of the method of dimensions

(1) *Checking formulae*
We use the fact that the two sides of a physical equation must have identical dimensions. Suppose that there is uncertainty as to whether the period T of a simple pendulum is either $2\pi\sqrt{l/g}$ or $2\pi\sqrt{g/l}$. Take the second case; dimensionally we find

$$T = \left(\frac{LT^{-2}}{L}\right)^{1/2} = \frac{1}{T}$$

This cannot be correct and hence the correct formula is $T = 2\pi\sqrt{l/g}$. The theory tells us nothing about π, which is the ratio of an arc to a radius. i.e. $[\pi] = L/L$ and π is dimensionless.

As another example, suppose that for torsional oscillations there is uncertainty as to whether the period $T = 2\pi\sqrt{I/C}$ or $T = 2\pi\sqrt{C/I}$, where I denotes moment of inertia, and C denotes torsional constant. Now $[I] = ML^2$ and $C = $ couple/unit angle of twist. Therefore $[C] = MLT^{-2}L$, i.e. since an angle has no dimensions C has the dimensions of a couple. Taking $T = 2\pi\sqrt{I/C}$, $T^2 = 4\pi I/C$ and dimensionally

$$T^2 = \frac{ML^2}{ML^2T^{-2}} = T^2$$

Hence the formula $T = 2\pi\sqrt{I/C}$ is correct.

(2) *Derivation of formulae*

Dimensional analysis is useful in cases where one is not sure how to deal with a problem theoretically, but has a good idea what the variables are. The theory gives the powers of the variables. For a formula involving three variables, there are three dimensions to play with, and one obtains three equations which may be solved for three unknowns.

6.2 Examples of uses of the method of dimensions

Pendulum formula

Consider the period of oscillation T of a simple pendulum. The variables to consider are m (mass of bob), l (length of pendulum), and g. Let $T = Km^x l^y g^z$, where K is a constant.

Dimensionally

$$T = K(M)^x (L)^y (LT^{-2})^z$$

Now the dimensions of the left-hand and right-hand sides must be the same. Thus the powers of M, L and T on both sides must individually be the same. Equating the powers of mass M, length L and time T we obtain respectively

$$0 = x$$
$$0 = y + z$$
$$1 = -2z$$

Thus $z = -\frac{1}{2}$, $y = \frac{1}{2}$ and $x = 0$, and so the period T is given by

$$T = Kl^{1/2} g^{-1/2} = K\sqrt{l/g}$$

The method of dimensions gives no information about the constant K, which is actually 2π.

Maximum velocity of a liquid flowing through a tube

The variables in this problem are the radius of the tube a, the viscosity of the liquid η, the pressure difference P, and the length of the tube l. There are thus four variables, but we can only expect three equations, one for each of M, L and T. Hence it is convenient to consider the pressure gradient P/l as a variable, rather than P and l separately.

Let the maximum velocity $v = K\eta^x a^y (P/l)^z$ where K is a constant. Dimensionally

$$LT^{-1} = K(ML^{-1}T^{-1})^x (L)^y (ML^{-2} T^{-2})^z$$

This equation gives

$$M \quad 0 = x + z$$
$$L \quad 1 = -x + y - 2z$$
$$T \quad -1 = -x - 2z$$

Adding the first and third equations we find $z = 1$. From the first equation we see that $x = -z = -1$, and from the second equation we have $y = 2$. Therefore

$$v = K \frac{a^2 P}{\eta l}$$

Detailed analysis gives $K = \frac{1}{4}$.

Poiseuille's equation
Here the variables are the radius of the tube a, the length of the tube l, the coefficient of viscosity of the fluid η, and the pressure difference P. Here we assume that the density is not required because the velocities are low. Let the volume transmitted per second be

$$V = K\eta^x a^y \left(\frac{P}{l}\right)^z$$

where K is a constant. Dimensionally

$$L^3 T^{-1} = K(ML^{-1}T^{-1})^x (L)^y (ML^{-2}T^{-2})^z$$

This equation gives

$$M \quad 0 = x + z$$
$$L \quad 3 = -x + y - 2z$$
$$T \quad -1 = -x - 2z$$

Therefore $z = 1$, $x = -1$, $y = 4$ and

$$V = K\frac{Pa^4}{\eta l}$$

Detailed analysis gives

$$V = \frac{\pi P a^4}{8\eta l}$$

Motion of a solid sphere through a fluid

Here the variables are the coefficient of viscosity of the fluid η, the terminal velocity of the sphere v, and the radius of the sphere a.

Let the frictional force $F = K\eta^x v^y a^z$, where K is a constant. Dimensionally

$$MLT^{-2} = K(ML^{-1}T^{-1})^x(LT^{-1})^y(L)^z$$

This equation gives

$$
\begin{array}{ll}
M & 1 = x \\
L & 1 = -x + y + z \\
T & -2 = -x - y
\end{array}
$$

Therefore $y = 1$, $z = 1$ and

$$F = K\eta v a$$

Detailed analysis gives $F = 6\pi\eta a v$ (Stokes' law).

The height of a Sessile drop

Suppose the height h (Fig. 6.1) depends on the surface tension of the liquid S, the density of the liquid ρ, and on g. Let $h = KS^x g^y \rho^z$, where K is a constant. Dimensionally

$$L = (MT^{-2})^x(LT^{-2})^y(ML^{-3})^z$$

This equation gives

$$
\begin{array}{lll}
M & 0 = x + z & \text{Therefore } z = -x \\
L & 1 = y - 3z & \\
T & 0 = -2x - 2y & \text{Therefore } y = -x
\end{array}
$$

To solve these simultaneous equations we substitute for z in the second equation using $z = -x$ from the first equation. We have

$$1 = y + 3x$$

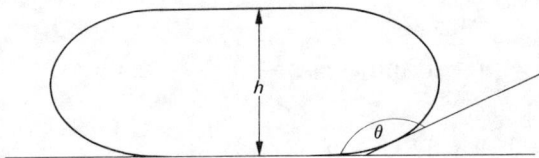

Fig. 6.1

and

$$0 = -2x - 2y$$

Multiplying the first equation by two and adding the two equations we have $x = \frac{1}{2}$. Thus $y = -\frac{1}{2}$ and $z = -\frac{1}{2}$. Therefore

$$h = KS^{1/2}g^{-1/2}\rho^{-1/2}$$

or

$$h^2 = \frac{KS}{\rho g}$$

Detailed analysis gives

$$h^2 = \frac{2S(1 - \cos \theta)}{\rho g}$$

where θ is the angle of contact.

Transition from streamline to turbulent flow

Dimensional analysis has been used in the theories of viscosity and aerodynamics in cases where four variables are involved. As an example of this, consider the case of a sphere of radius a, falling through a fluid of density ρ and coefficient of viscosity η, with a terminal velocity v sufficiently high that turbulence might be expected to occur. In these circumstances the retarding force F might be supposed to be a function of a, η, ρ, and v. Thus, let $F = Ka^x\eta^y\rho^zv^n$. Dimensionally

$$MLT^{-2} = L^x(ML^{-1}T^{-1})^y(ML^{-3})^z(LT^{-1})^n$$

This equation gives

$$
\begin{array}{ll}
M & 1 = y + z \\
L & 1 = x - y - 3z + n \\
T & -2 = -y - n
\end{array}
$$

Regarding n as an unknown we have

$$y = 1 - z$$
$$-2 = -1 + z - n$$

and thus $z = n - 1$ and $y = 2 - n$.
Also

$$x = 1 + y + 3z - n$$
$$= 1 + 2 - n + 3n - 3 - n$$

Therefore $x = n$. Thus, leaving n unknown, we may write

$$F = Ka^n\eta^{2-n}\rho^{n-1}v^n$$

and collecting everything to the power n together we have

$$F = \frac{K\eta^2}{\rho}\left(\frac{a\rho v}{\eta}\right)^n$$

This formula is dimensionally correct whatever the value of n, and thus $a\rho v/\eta$ must be dimensionless. If experiment shows that force F is proportional to velocity, then one can put $n = 1$ in this formula giving

$$F = \frac{K\eta^2}{\rho}\left(\frac{a\rho v}{\eta}\right)$$
$$= K\eta av$$

K is found to have a value 6π, i.e. $F = 6\pi\eta av$, and Stokes' law holds (streamline conditions). It is always found experimentally that force is proportional to velocity when $a\rho v/\eta$ is small ($a\rho v/\eta < 1$).

If experiment shows that force F is proportional to (velocity)^2, then $n = 2$ and

$$F = \text{constant}\ \frac{\eta^2}{\rho}\left(\frac{a\rho v}{\eta}\right)^2 = \text{constant}\ a^2\rho v^2$$

It is always found experimentally that force is proportional to (velocity)^2 when $a\rho v/\eta$ is large. Thus at some particular value of $a\rho v/\eta$ there is a change from streamline to turbulent conditions. In many fluid flow problems the dimensionless product $(\rho/\eta) \times$ a length $\times$ a velocity occurs. It is called the Reynolds number for the particular problem, and its value gives an indication as to whether flow is streamline or turbulent.

In the case of flow through a tube of radius a, the mean fluid velocity at which turbulence occurs v_c, may be assumed to depend on the coefficient of viscosity of the fluid η, its density ρ and on a. Let $v_c = K\rho^x\eta^ya^z$, where K is a constant. Dimensionally

$$LT^{-1} = K(ML^{-3})^x(ML^{-1}T^{-1})^y(L)^z$$

This equation gives

$$M \quad 0 = x + y$$
$$L \quad 1 = -3x - y + z$$
$$T \quad -1 = -y$$

Therefore $x = -1$, $y = 1$, $z = -1$ and so

$$v_c = K\frac{\eta}{a\rho}$$

Hence a high viscosity increases v_c, and encourages streamline flow, while a high density does the reverse. Reynolds obtained this relation experimentally for narrow tubes, and $K = a\rho v_c/\eta$ Reynolds' number, has a value of approximately 1000. If Reynolds' number is less than about 1000 streamline or laminar flow occurs. Substituting for v_c using

$$v_c = \frac{a^2 P}{4\eta L}$$

(actually the formula for the maximum velocity in the tube) where P is the pressure difference across the tube, and L is the length of the tube, we have

$$K = \frac{a\rho}{\eta}\frac{a^2 P}{4\eta L} = \frac{1}{4}\frac{a^3}{\eta^2}\frac{P}{L}\rho$$

Hence the use of tubes of large radius and the use of high pressure gradients is to be avoided, if streamline conditions are to obtain.

7 Vibrations

We will deal with some of the problems which arise in vibrations by using complex numbers, but, before commencing, we will give a brief outline of the mathematics required. The reasons for using complex numbers are as follows: every student covers complex numbers in his first year mathematics course; also, the use of complex numbers is essential in alternating current theory; they are also very useful in physical optics. Thus, since the method is very elegant and very powerful, and since the student will use the method elsewhere in any case, we will use it to some extent in dealing with vibrations.

With reference to Fig. 7.1, a complex number z, may be written $z = x + iy$, where x is called its real part, and y is called its imaginary part, and where $i = \sqrt{-1}$. But $x = r \cos \theta$ and $y = r \sin \theta$. Thus

$$x + iy = r \cos \theta + ir \sin \theta = r(\cos \theta + i \sin \theta)$$

which is called the polar form of the complex number. Clearly $r = \sqrt{x^2 + y^2}$, and is called the modulus of the complex number, written $|z|$. Tan $\theta = y/x$, and θ is called the argument of the complex number.

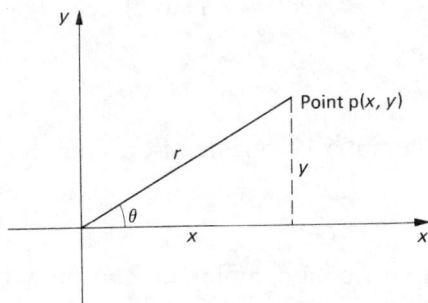

Fig. 7.1

Now, both sin θ and cos θ can be written as power series, i.e.

$$\sin \theta = \theta - \frac{\theta^3}{3!} + \frac{\theta^5}{5!} - \cdots$$

and

$$\cos \theta = 1 - \frac{\theta^2}{2!} + \frac{\theta^4}{4!} - \cdots$$

But

$$e^x = 1 + x + \frac{x^2}{2!} + \frac{x^3}{3!} + \cdots$$

Therefore

$$e^{i\theta} = 1 + i\theta + \frac{(i\theta)^2}{2!} + \frac{(i\theta)^3}{3!} + \frac{(i\theta)^4}{4!} + \cdots$$

and using the fact that $i^2 = -1$, $i^3 = -i$, $i^4 = 1$, etc. we may write

$$e^{i\theta} = 1 + i\theta - \frac{\theta^2}{2!} - i\frac{\theta^3}{3!} + \frac{\theta^4}{4!} + i\frac{\theta^5}{5!} \cdots$$

$$= 1 - \frac{\theta^2}{2!} + \frac{\theta^4}{4!} \cdots + i\left(\theta - \frac{\theta^3}{3!} + \frac{\theta^5}{5!} \cdots\right)$$

$$= \cos \theta + i \sin \theta$$

(by comparison with the power series for cos θ and sin θ).
Thus

$$e^{i\theta} = \cos \theta + i \sin \theta$$

and the complex number

$$z = x + iy = r(\cos \theta + i \sin \theta)$$

may be written

$$z = re^{i\theta}$$

$\bar{z}$ is called the complex conjugate of the complex number z, and is given by

$$\bar{z} = x - iy = re^{-i\theta}$$

However complicated a complex number may be, its complex conjugate is always obtained by replacing i by $-i$, wherever it appears. Note also that $z\bar{z} = re^{i\theta}.re^{-i\theta} = r^2$.

7.1 Representation of simple harmonic vibration by complex numbers

Now, for a particle undergoing simple harmonic motion, the displacement

$$y = a \sin (\omega t + \delta) \quad \text{or} \quad y = a \cos (\omega t + \delta)$$

(either sine or cosine may be used), where a is the amplitude of vibration, ω is the angular frequency, related to the period T by $T = 2\pi/\omega$, and δ is the phase constant. $\omega t + \delta$ is the phase. However, from what we have said about complex numbers, we can represent the simple harmonic vibration by

$$y = a e^{i(\omega t + \delta)},$$

where y is now a complex number. Thus

$$y = a \cos (\omega t + \delta) + ia \sin (\omega t + \delta)$$

and either the real or the imaginary part of y, gives the displacement of the simple harmonic vibration. $y\bar{y}$ ($\bar{y}$ is the complex conjugate of y) gives the square of the amplitude a^2.

7.2 Compounding of two simple harmonic vibrations

Vibrations of the same frequency with a phase difference between them
This is the problem of optical interference. Let the vibrations be represented in complex terms by

$$y_1 = a e^{i\omega t}$$

and

$$y_2 = b e^{i(\omega t + \delta)}$$

(different amplitudes in general). δ is the phase difference between the vibrations, which have the same frequency (ω the same in both cases).

By the principle of superposition, we may simply add to obtain the resultant vibration. Thus, consider

$$Y = y_1 + y_2 = e^{i\omega t}\left[a + b e^{i\delta}\right]$$

Now $a + b e^{i\delta}$ is complex, with a real part $a + b \cos \delta$, and an imaginary part $b \sin \delta$. Thus it may be written $r e^{i\theta}$, where its

modulus

$$r = \sqrt{a^2 + b^2 + 2ab \cos \delta}$$

and

$$\tan \theta = \frac{b \sin \delta}{a + b \cos \delta}$$

Thus

$$Y = e^{i\omega t} r e^{i\theta}$$

$$= \sqrt{a^2 + b^2 + 2ab \cos \delta} \exp\left[i \tan^{-1}\left(\frac{b \sin \delta}{a + b \cos \delta}\right)\right] e^{i\omega t}$$

$$= \sqrt{a^2 + b^2 + 2ab \cos \delta}\; e^{i(\omega t + \theta)}$$

and this represents the resultant vibration in complex terms. The resultant vibration has an amplitude

$$\sqrt{a^2 + b^2 + 2ab \cos \delta}$$

and a phase difference with respect to y_1 of θ, where

$$\tan \theta = \frac{b \sin \delta}{a + b \cos \delta}$$

The amplitude will be a maximum equal to $(a + b)$ when $\cos \delta = 1$, i.e.

$$\delta = 0, 2\pi, 4\pi, \ldots$$

and a minimum equal to $(a - b)$ when $\cos \delta = -1$, i.e.

$$\delta = \pi, 3\pi, 5\pi, \ldots$$

If the amplitudes a and b are the same, the resultant amplitude will be twice the amplitude of either of the constituent vibrations, at a maximum, and zero at a minimum.

Fig. 7.2 represents the solution of this problem by the use of a vector diagram.

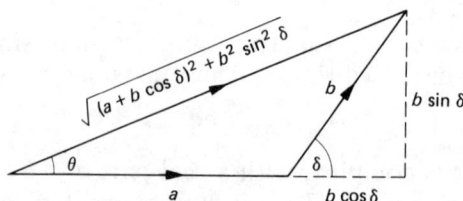

Fig. 7.2

Vibrations of different frequencies

Let the vibrations be represented in complex terms by

$$y_1 = ae^{i\omega t}$$

and

$$y_2 = be^{i(\omega + \omega')t}$$

$\omega'/2\pi$ is the frequency difference. As before, the displacement of the resultant vibration

$$Y = y_1 + y_2 = e^{i\omega t}\left[a + be^{i\omega't}\right]$$

But $a + be^{i\omega't}$ is complex, and may be written $re^{i\theta}$, where

$$r = \sqrt{a^2 + b^2 + 2ab \cos \omega't}$$

and where

$$\tan \theta = \frac{b \sin \omega't}{a + b \cos \omega't}$$

Thus

$$Y = \sqrt{a^2 + b^2 + 2ab \cos \omega't} \; e^{i(\omega t + \theta)}$$

and we see that the amplitude of the resultant vibration is a function of time. So also is the phase constant θ. However, if the frequency difference is very small (i.e. ω' is small), then θ is small, and the vibration approximates to $Y = Ae^{i\omega t}$, where A, the amplitude of the resultant vibration, equals

$$\sqrt{a^2 + b^2 + 2ab \cos \omega't}$$

The resultant amplitude varies with time with a period $2\pi/\omega'$. This is the phenomenon of 'beats', and $\omega'/2\pi$ is called the beat frequency. The beating is much more violent if $a = b$, and the resultant amplitude then goes right down to zero.

Fig. 7.3 illustrates the variation in amplitude of the resultant vibration which occurs when two vibrations of nearly the same frequency are compounded.

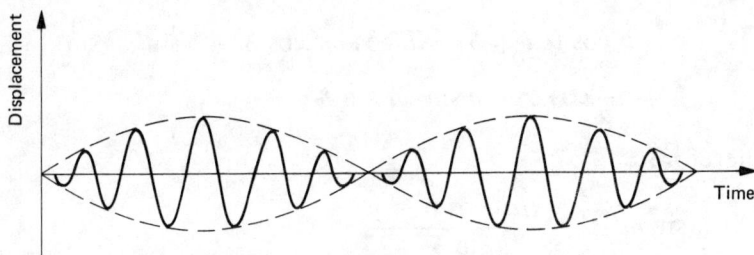

Fig. 7.3

7.3   Lissajous' figures

Consider a particle acted upon by two simple harmonic forces of the same frequency, acting at right angles to one another (Fig. 7.4). We may write for the displacement in the x direction $x = a \cos \omega t$, and for the displacement in the y direction $y = b \cos (\omega t + \delta)$, where $\omega/2\pi$ is the frequency of the vibrations, and δ is the phase difference between them. x and y are not complex now.

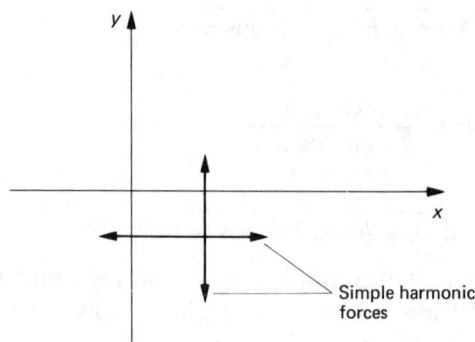

Fig. 7.4

If time t can be eliminated from these two equations, a closed curve in the xy plane is obtained. In fact t can be eliminated when the frequencies are the same, and even if the frequencies are not the same, t can be eliminated if the frequency ratio is that of two whole numbers, i.e. the frequencies are commensurate. Now

$$x = a \cos \omega t$$

and

$$y = b \cos (\omega t + \delta) = b \cos \omega t \cos \delta - b \sin \omega t \sin \delta$$

$$= b \frac{x}{a} \cos \delta - b \sin \omega t \sin \delta$$

Therefore

$$\sin \omega t = \frac{b(x/a) \cos \delta - y}{b \sin \delta}$$

and

$$\sin^2 \omega t = \frac{(b^2x^2/a^2) \cos^2 \delta + y^2 - 2(bxy/a) \cos \delta}{b^2 \sin^2 \delta}$$

$$\cos^2 \omega t = x^2/a^2$$

Multiplying these last two equations by $a^2b^2 \sin^2 \delta$, and then adding them, we have

$$b^2x^2 \sin^2 \delta + b^2x^2 \cos^2 \delta + a^2y^2 - 2abxy \cos \delta = a^2b^2 \sin^2 \delta$$

that is

$$b^2x^2 + a^2y^2 - 2abxy \cos \delta = a^2b^2 \sin^2 \delta$$

Dividing by a^2b^2, we have

$$\frac{x^2}{a^2} + \frac{y^2}{b^2} - \frac{2xy}{ab} \cos \delta = \sin^2 \delta$$

This is the equation of a curve completely contained in the rectangle $2a$ by $2b$. With reference to Fig. 7.5, if $\delta = 0$, then

$$\left(\frac{x}{a} - \frac{y}{b}\right)^2 = 0$$

which is the equation of two straight lines which have come together, and $y = bx/a$.

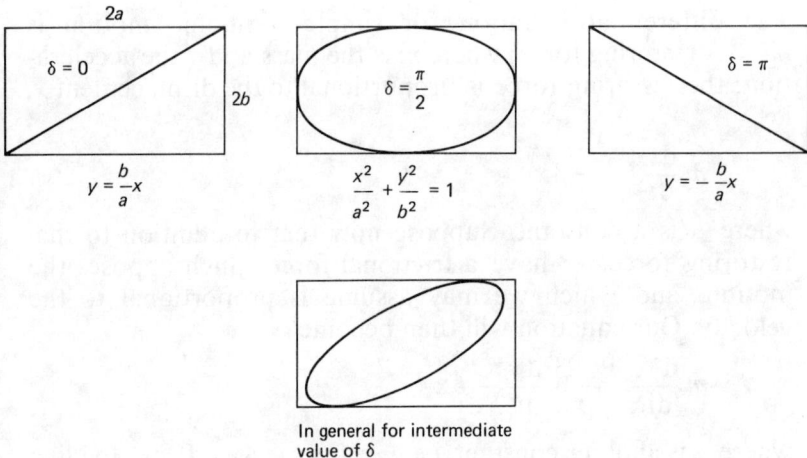

In general for intermediate value of δ

Fig. 7.5

If $\delta = \pi$, then

$$\left(\frac{x}{a} + \frac{y}{b}\right)^2 = 0 \text{ and } y = -bx/a$$

(equation of line with negative slope). If $\delta = \pi/2$, then

$$\frac{x^2}{a^2} + \frac{y^2}{b^2} = 1$$

which is the equation of an ellipse with the origin at its centre. If in addition $a = b = r$, we have $x^2 + y^2 = r^2$, which is the equation of a circle. For intermediate values of δ, we get ellipses which are tilted with respect to the ellipse obtained when $\delta = \pi/2$. If the frequencies are different, but are not commensurate, i.e. their ratio is not that of two whole numbers, no stationary figure is obtained. If the frequency ratio is 2/1, and $\delta = \pi/2$, we get a figure of eight. If the frequencies are nearly the same, the pattern passes slowly through all the forms, from straight line through various ellipses back to straight line again. Lissajous' figures are best displayed using a cathode ray oscilloscope, with an alternating voltage on each of the X and Y plates.

7.4   Damped harmonic motion

The differential equation for simple harmonic motion is $mf = -$restoring force, where m is the mass and f the acceleration; the restoring force is proportional to the displacement x, i.e.

$$m\frac{\mathrm{d}^2x}{\mathrm{d}t^2} = -\beta x$$

where β is a constant. Suppose now that in addition to the restoring force we have a frictional force which opposes the motion, and which we may assume is proportional to the velocity. Our equation will then become

$$m\frac{\mathrm{d}^2x}{\mathrm{d}t^2} = -\alpha\frac{\mathrm{d}x}{\mathrm{d}t} - \beta x$$

where α is another constant, i.e. $m\ddot{x} + \alpha\dot{x} + \beta x = 0$, or, dividing by m,

$$\ddot{x} + 2k\dot{x} + \omega^2 x = 0$$

where $2k = \alpha/m$ and $\omega^2 = \beta/m$. ω^2 is the restoring force per unit mass, per unit displacement. The standard method of solving a differential equation like this is to assume a solution of the form $x = Ae^{\lambda t}$, where A is a constant, and λ is another constant which must be chosen to satisfy the differential equation. Thus

$$x = Ae^{\lambda t}$$

and differentiating,

$$\dot{x} = A\lambda e^{\lambda t}$$

Differentiating again

$$\ddot{x} = A\lambda^2 e^{\lambda t}$$

Substituting these values in the differential equation we have

$$A\lambda^2 e^{\lambda t} + 2kA\lambda e^{\lambda t} + \omega^2 Ae^{\lambda t} = 0$$

i.e.

$$\lambda^2 + 2k\lambda + \omega^2 = 0$$

and $Ae^{\lambda t}$ will be a solution of the differential equation if λ is a root of this quadratic, i.e. if

$$\lambda = \frac{-2k \pm \sqrt{4k^2 - 4\omega^2}}{2} = -k \pm \sqrt{k^2 - \omega^2}$$

which gives two values of λ, thus

$$x = A \exp\left[(-k + \sqrt{k^2 - \omega^2})t\right]$$

is one solution, and

$$x = B \exp\left[(-k - \sqrt{k^2 - \omega^2})t\right]$$

is another, where A and B are arbitrary constants.

Now the solution of a second-order differential equation, such as the one we are considering, must contain two arbitrary constants, and the general solution is obtained by adding the solutions we have just found. For example, consider the acceleration due to gravity: $d^2x/dt^2 = g$.

Therefore, integrating

$$\frac{dx}{dt} = gt + C$$

and integrating again

$$x = \tfrac{1}{2}gt^2 + Ct + D$$

where C and D are arbitrary constants, which are determined from additional information about the problem (e.g. values of dx/dt and x when $t = 0$). Thus the general solution of our differential equation for damped harmonic motion is

$$x = e^{-kt}\left[A\exp(\sqrt{k^2 - \omega^2}t) + B\exp(-\sqrt{k^2 - \omega^2}t)\right]$$

e^{-kt} is the decay term, and the behaviour of the term in square brackets depends upon whether k is greater or less than the natural angular frequency ω.

We concentrate on the behaviour of

$$A\exp(\sqrt{k^2 - \omega^2}t) + B\exp(-\sqrt{k^2 - \omega^2}t) = u \quad \text{say}$$

Let

$$x = e^{-kt}u$$

Differentiating

$$\dot{x} = -ke^{-kt}u + \dot{u}e^{-kt}$$

Differentiating again

$$\ddot{x} = k^2 e^{-kt}u - \dot{u}ke^{-kt} - \dot{u}ke^{-kt} + \ddot{u}e^{-kt}$$
$$= k^2 e^{-kt}u - 2\dot{u}ke^{-kt} + \ddot{u}e^{-kt}$$

Putting these values into the equation

$$\ddot{x} + 2k\dot{x} + \omega^2 x = 0$$

we have

$$k^2 u - 2k\dot{u} + \ddot{u} - 2k^2 u + 2k\dot{u} + \omega^2 u = 0$$

Therefore

$$\ddot{u} + (\omega^2 - k^2)u = 0$$

If $k > \omega$ (large damping since damping depends on k), then $(\omega^2 - k^2)$ is negative, and

$$\ddot{u} - (k^2 - \omega^2)u = 0$$

In this case the solution must be left in its original form, i.e.:

$$u = A\exp(\sqrt{k^2 - \omega^2}t) + B\exp(-\sqrt{k^2 - \omega^2}t)$$

and

$$x = e^{-kt}\left[A \exp(\sqrt{k^2 - \omega^2}t) + B \exp(-\sqrt{k^2 - \omega^2}t)\right]$$

If $k < \omega$ (small damping), then $(\omega^2 - k^2)$, is positive, and

$$\ddot{u} = -(\omega^2 - k^2)u$$

This is the differential equation for ordinary simple harmonic motion, and

$$u = a \sin(\sqrt{\omega^2 - k^2}t + \delta)$$

where a and δ are arbitrary constants. Thus

$$x = e^{-kt}a \sin(\sqrt{\omega^2 - k^2}t + \delta)$$

This equation looks similar to the equation for simple harmonic motion because of the sine term. However, the motion is not simple harmonic because the amplitude ae^{-kt} decreases as t increases.

If $k = \omega$ (critical damping), then $\ddot{u} = 0$, and integrating gives $\dot{u} = C$. Integrating again, we have

$$u = Ct + D$$

where C and D are arbitrary constants. Thus

$$x = e^{-kt}(Ct + D)$$

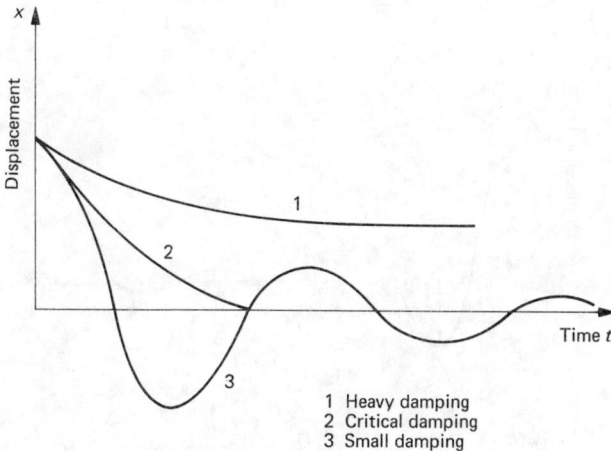

1 Heavy damping
2 Critical damping
3 Small damping

Fig. 7.6

With reference to Fig. 7.6, curve 1 is for heavy damping, $k >$ ω, i.e. the frictional opposition to the motion is great in comparison with the restoring force, and

$$x = e^{-kt}\left[A\exp(\sqrt{k^2 - \omega^2}t) + B\exp(-\sqrt{k^2 - \omega^2}t)\right]$$

In this case the displacement decreases slowly, according to an exponential equation, and never becomes negative.

Curve 2 is for critical damping, $k = \omega$, where oscillations just fail to occur, and where the system settles to its equilibrium position in the shortest possible time. In this case

$$x = e^{-kt}(Ct + D)$$

Curve 3 is for small damping, $k < \omega$, and the system oscillates according to the equation

$$x = ae^{-kt}\sin(\sqrt{\omega^2 - k^2}t + \delta)$$

In this case the amplitude of oscillation decays exponentially, and the displacement gradually falls to zero. Since the period of oscillation is $2\pi/\sqrt{\omega^2 - k^2}$, as compared with $2\pi/\omega$ when there is no damping, damping gives a lower frequency of oscillation.

Consider $x = ae^{-kt}\sin(\sqrt{\omega^2 - k^2}t + \delta)$ (small damping and $k < \omega$); thus if $\delta = 0$

$$x = ae^{-kt}\sin\left(2\pi\frac{t}{T}\right)$$

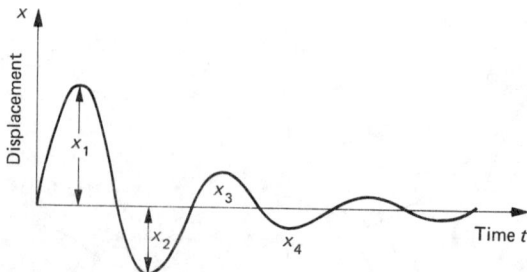

Fig. 7.7

(Fig. 7.7) where T, the period of oscillation, is $2\pi/\sqrt{\omega^2 - k^2}$ (e.g. a physical system which has a velocity at $t = 0$ but zero displacement).

At $t = 0$ $\quad x = 0$

At $t = T/4$ $\quad x_1 = a\mathrm{e}^{-kT/4}$

At $t = 3T/4$ $\quad x_2 = -a\mathrm{e}^{-3kT/4}$

Then numerically (ignoring any minus signs),

$$\frac{x_1}{x_2} = \frac{x_2}{x_3} = \frac{x_3}{x_4} \ldots = \mathrm{e}^{kT/2} = d,$$

where d, the ratio of successive amplitudes, is called the decrement, and $\log_e d = \lambda = kT/2$, is called the logarithmic decrement.

7.5 Discharge of a capacitor through an inductance and a resistance

As an example of damped harmonic motion we will consider the discharge of a capacitor of capacitance C, through an inductance L and a resistance R (Fig. 7.8). The capacitor is

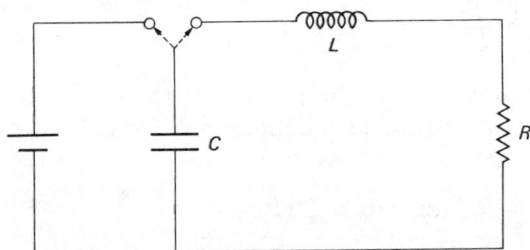

Fig. 7.8

charged by means of the battery, and is then suddenly discharged at time $t = 0$. At time t, let the charge on the capacitor be $q(t)$, let the current be $i(t)$, and let us assume that the current is flowing in a clockwise sense, i.e. that the capacitor is discharging. $q(t)/C$ is the voltage across the capacitor, and since L is defined as flux per unit current, $L\,\mathrm{d}i/\mathrm{d}t$ will be the e.m.f. induced in the inductance (Faraday's law).

Now Kirchhoff's second law can be applied to the circuit at

any instant of time. Thus

$$\frac{q(t)}{C} - L\frac{di}{dt} = Ri$$

(the negative sign is due to Lenz's law). But

$$i = -dq/dt$$

(q is decreasing with time). Therefore

$$di/dt = -d^2q/dt^2$$

Thus

$$L\frac{d^2q}{dt^2} + R\frac{dq}{dt} + \frac{q}{C} = 0$$

that is

$$\frac{d^2q}{dt^2} + 2k\frac{dq}{dt} + \omega^2 q = 0$$

where $2k = R/L$ and $\omega^2 = 1/LC$. This we recognize as the equation for damped harmonic motion, and the solution will be harmonically varying (oscillatory discharge), if $\omega^2 - k^2 > 0$, i.e. if

$$\frac{1}{LC} > \frac{R^2}{4L^2} \quad \text{or} \quad \frac{4L}{C} > R^2$$

In the limit when the discharge just ceases to be oscillatory, for a given L/C, a resistance is defined, i.e. for critical damping

$$\frac{4L}{C} = R_c^2 \quad \text{or} \quad R_c = 2\sqrt{\frac{L}{C}}$$

where R_c is called the critical damping resistance. If the resistance is larger than R_c, the discharge is exponential, and if it is smaller than R_c it is oscillatory. In the case of an oscillatory discharge, the frequency of oscillation is

$$\frac{\sqrt{\omega^2 - k^2}}{2\pi} = \frac{1}{2\pi}\sqrt{\frac{1}{LC} - \frac{R^2}{4L^2}}$$

and the logarithmic decrement λ, which indicates how rapidly

the amplitude decays, is given by

$$\lambda = \tfrac{1}{2}kT = \frac{k}{2}\frac{2\pi}{\sqrt{\omega^2 - k^2}}$$

$$= \frac{R}{4L}\frac{2\pi}{\sqrt{(1/LC) - (R^2/4L^2)}}\ .$$

Thus λ increases as R increases.

If R is zero, the damped harmonic motion differential equation becomes

$$L\frac{d^2q}{dt^2} + \frac{q}{C} = 0$$

that is

$$\frac{d^2q}{dt^2} = \frac{-q}{LC}$$

which represents simple harmonic motion. Thus in this case $q = q_0 \sin \omega t$, where $\omega = 1/\sqrt{LC}$ and q_0 is the initial charge on the condenser. The frequency of oscillation $1/2\pi\sqrt{LC}$ is called the resonant frequency of the circuit. However, in practice R is never zero, since neither the inductance nor the capacitor can be completely pure (each will have some resistance).

7.6 Oscillations in the tuned grid circuit of a triode valve

In Fig. 7.9, R is the grid circuit resistance, C the grid capacitance, and M the mutual inductance between the anode and

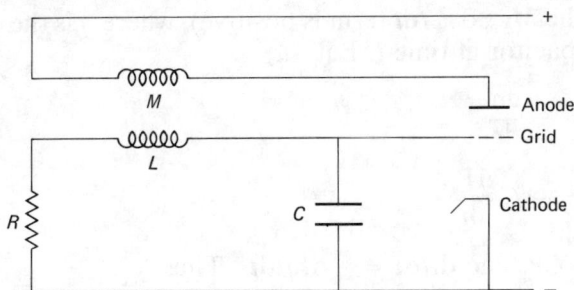

Fig. 7.9

grid coils. Consider the *LCR* circuit. The e.m.f. due to *M* is di_a/dt, where i_a is the anode current at any instant, and the *M* mutual conductance of the valve g_m is defined as

$$g_m = \left(\frac{\partial i_a}{\partial V_g}\right)_{V_a},$$

(i.e. rate of change of anode current with grid voltage V_g, when the anode voltage V_a is maintained constant). But provided that the valve is operating in the linear regions of the i_a, V_g curves, g_m will be a constant independent of V_a (the i_a, V_g curves for different V_a, are straight parallel lines over a considerable region) and

$$g_m = \frac{di_a}{dV_g}$$

$$= \frac{di_a}{dt}\frac{dt}{dV_g}.$$

Therefore the e.m.f. induced in the *LCR* circuit due to M is given by

$$Mg_m \frac{dV_g}{dt}$$

Now, assuming an anticlockwise flow of current at any instant (capacitor discharging), and applying Kirchhoff's second law to the *LCR* circuit, we may write

$$\frac{q(t)}{C} - L\frac{di}{dt} + Mg_m\frac{dV_g}{dt} = Ri$$

(assume the $Mg_m \, dV_g/dt$ term is positive), where q is the charge on the capacitor at time t. But

$$i = -\frac{dq}{dt}$$

$$= -C\frac{dV_g}{dt}$$

since $q = CV_g$ and $di/dt = -d^2q/dt^2$. Thus

$$L\frac{d^2q}{dt^2} + \left(R + \frac{Mg_m}{C}\right)\frac{dq}{dt} + \frac{q}{C} = 0$$

that is

$$\frac{d^2q}{dt^2} + 2k\frac{dq}{dt} + \omega^2 q = 0,$$

where

$$2k = \frac{R + (Mg_m/C)}{L}$$

and $\quad \omega^2 = \dfrac{1}{LC}$

This is the equation for damped harmonic motion, and $[R + (Mg_m/C)]/L$ is the damping term. Now g_m and C are both positive, thus if M is negative (primary and secondary voltages $180°$ out of phase) and $R = Mg_m/C$, the damping term vanishes. We then have

$$L\frac{d^2q}{dt^2} = -\frac{q}{C}$$

This is the equation for simple harmonic motion. Thus in this case

$$q = q_0 \sin(\omega t + \delta)$$

where $\omega = 1/\sqrt{LC}$, and oscillations of constant amplitude continue in the circuit.

In general

$$\ddot{q} + 2k\dot{q} + \omega^2 q = 0$$

and if $k < \omega$

$$q = q_0 e^{-kt} \sin(\sqrt{\omega^2 - k^2}\, t + \delta)$$

But

$$k = \frac{R + (Mg_m/C)}{2L},$$

and g_m and C are both positive. Thus if M is negative and $Mg_m > RC$, k will be negative. Clearly in this case the oscillations will build up (positive exponential term), the oscillations growing exponentially with time (Fig. 7.10). On the other hand if $k < \omega$ and M is positive, k will be positive, and the oscillations

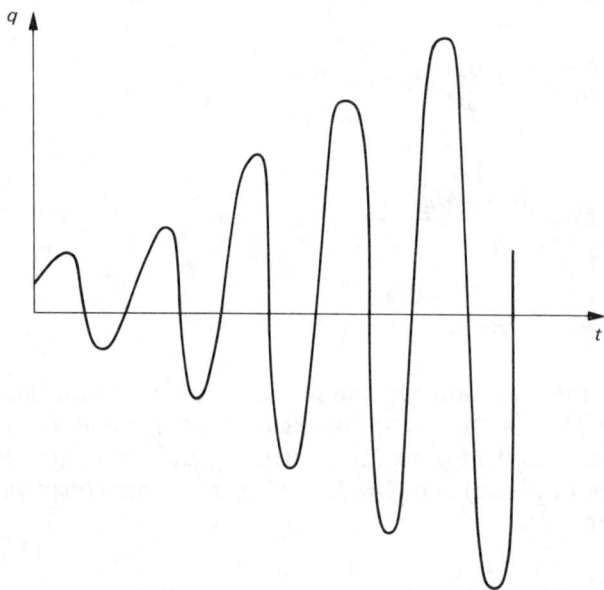

Fig. 7.10

will die away exponentially. Of course if $k > \omega$, no oscillations at all will occur in the circuit.

Thus oscillations will be maintained or will build up if

$$M \geqq RC/g_m$$

In practice if the condition for a build up of oscillations is satisfied, the oscillations do not increase indefinitely but become anharmonic. The reason for this is that if the swing of grid voltage V_g becomes too great, g_m will no longer be constant and our theory will no longer be valid. Also if the grid becomes too negative, the anode current will be cut off, while if it becomes too positive, it will draw current.

7.7   Forced vibrations

When considering damped harmonic motion, we had the equation

$$m\frac{d^2x}{dt^2} = -\alpha\frac{dx}{dt} - \beta x$$

that is

$$m\ddot{x} + \alpha\dot{x} + \beta x = 0$$

where m denotes mass, $\alpha\dot{x}$ is the damping term due to frictional resistance, and β is the restoring force per unit displacement. Suppose now that in addition, an harmonically varying force $F \cos pt$ (frequency $p/2\pi$) is applied to the system. Our equation then becomes

$$m\ddot{x} + \alpha\dot{x} + \beta x = F \cos pt$$

or, dividing by m

$$\ddot{x} + 2k\dot{x} + \omega^2 x = a \cos pt$$

where $2k = \alpha/m$, $\omega^2 = \beta/m$ and $a = F/m$

Assume that the steady-state solution is harmonic, and is given by $x = A \cos(nt + f)$, where A and f are constants to be determined. We are assuming then that the system will oscillate with a frequency $n/2\pi$ and that it will have some phase constant f. We now write our forced vibration equation in complex form, i.e.

$$\ddot{z} + 2k\dot{z} + \omega^2 z = a e^{ipt}$$

where z is complex. But

$$z = A e^{i(nt+f)}$$

(our suggested solution in complex form, x is the real part of z). Therefore

$$\dot{z} = inz \quad \text{and} \quad \ddot{z} = i^2 n^2 z = -n^2 z$$

Substituting these values in our complex equation, we have

$$(-n^2 + 2kin + \omega^2)z = a e^{ipt}$$

Therefore

$$z = \frac{a e^{ipt}}{\omega^2 - n^2 + 2kni}$$

and multiplying top and bottom by the complex conjugate of the denominator, we have

$$z = \frac{(\omega^2 - n^2 - 2kni)a e^{ipt}}{(\omega^2 - n^2)^2 + 4k^2 n^2}$$

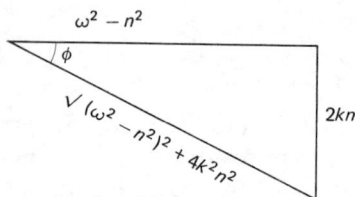

Fig. 7.11

But $(\omega^2 - n^2) - 2kni$ is complex, and may be written
$$\sqrt{(\omega^2 - n^2)^2 + 4k^2n^2}\,e^{-i\phi},$$
where $\tan \phi = 2kn/(\omega^2 - n^2)$ (Fig. 7.11). Thus
$$z = \frac{a\,e^{i(pt-\phi)}}{\left[(\omega^2 - n^2)^2 + 4k^2n^2\right]^{1/2}}$$
and our solution x is the real part of this, that is
$$x = \frac{a\,\cos\,(pt - \phi)}{\left[(\omega^2 - n^2)^2 + 4k^2n^2\right]^{1/2}}$$
where $\tan \phi = 2kn/(\omega^2 - n^2)$. Comparing this with our assumption
$$x = A \cos (nt + f)$$
we see that $n = p$, $f = -\phi$ and
$$A = \frac{a}{\left[(\omega^2 - p^2)^2 + 4k^2p^2\right]^{1/2}}$$

This solution is called the particular integral, but the complete solution also includes the complementary function which is the solution of the equation $\ddot{x} + 2k\dot{x} + \omega^2x = 0$. This, as we have seen, gives a function for x which decreases to zero as t increases, due to the presence of the term e^{-kt}.

Thus the system will execute undamped oscillations of frequency equal to that of the driving oscillation, and with a phase lag ϕ with respect to the driving oscillation. Looking at the expression for the amplitude of oscillation A, we see that the greater the difference between ω and p, the smaller will be the amplitude. If k is very small, the amplitude has a maximum when $\omega = p$ (resonance).

Fig. 7.12 shows how the amplitude varies with the angular

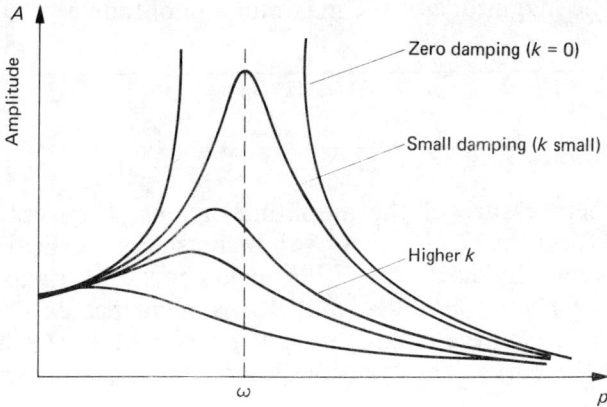

Fig. 7.12

frequency of the driving oscillation p. The smaller the damping the sharper the peak, and, as the damping increases, the peak moves towards lower values of p. When $p \gg \omega$ the amplitude tends to zero irrespective of the amount of damping. To obtain the value of p which gives maximum amplitude, we differentiate A with respect to p. Thus

$$A = \frac{a}{\left[(\omega^2 - p^2)^2 + 4k^2p^2\right]^{1/2}}$$

and

$$\frac{\mathrm{d}A}{\mathrm{d}p} = \frac{-\frac{1}{2}a(4p^3 - 4\omega^2p + 8k^2p)}{\left[(\omega^2 - p^2)^2 + 4k^2p^2\right]^{3/2}}$$

For a maximum $\mathrm{d}A/\mathrm{d}p = 0$, i.e.

$$4p^3 - 4\omega^2p + 8k^2p = 0$$

or

$$p^2 = \omega^2 - 2k^2$$

i.e.

$$p = \sqrt{\omega^2 - 2k^2}$$

and the frequency at maximum amplitude is $\sqrt{\omega^2 - 2k^2}/2\pi$. This compares with a frequency $\sqrt{\omega^2 - k^2}/2\pi$ for damped oscillations, and a frequency $\omega/2\pi$ for natural undamped oscillations. The magnitude of the maximum amplitude is

$$\frac{a}{\left[(\omega^2 - \omega^2 + 2k^2)^2 + 4k^2(\omega^2 - 2k^2)\right]^{1/2}}$$

$$= \frac{a}{(4k^4 + 4k^2\omega^2 - 8k^4)^{1/2}}$$

Another feature of the amplitude against p curves to be noticed from Fig. 7.12 is that, with increasing k, the peak in the curve eventually disappears. This occurs because for maximum amplitude $p = \sqrt{\omega^2 - 2k^2}$, and if k is such that $2k^2 = \omega^2$ or $k = \omega/\sqrt{2}$, then the maximum is at $p = 0$ and there is no real peak in the curve. Thus the condition for no peak in the curve is $k \geqq \omega/\sqrt{2}$.

The phase of the forced vibrations

Our solution of the forced vibration equation is $x = A \cos (pt - \phi)$, where the phase lag ϕ with respect to the forcing oscillation is given by

$$\tan \phi = \frac{2kp}{\omega^2 - p^2}$$

Thus

$$\sin \phi = \frac{2kp}{\left[4k^2p^2 + (\omega^2 - p^2)^2\right]^{1/2}}$$

ϕ lies between two limiting values 0 and π, since $\sin \phi$ is always positive. Suppose $p \gg \omega$, then $\tan \phi \rightarrow -2k/p \rightarrow -0$ for small k, i.e. $\phi \rightarrow \pi$ (resultant lags in phase by 180° with respect to the driver).

Since $\tan \phi$ is negative for $p > \omega$, ϕ must in this case lie between $\pi/2$ and π (Fig. 7.13).

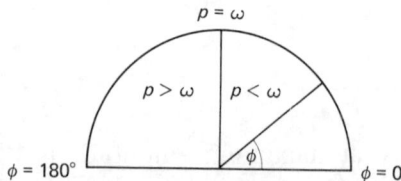

Fig. 7.13

On the other hand if $p \ll \omega$, then $\tan \phi \to 2kp/\omega^2 \to 0$ for small k or small p, i.e. $\phi \to 0$ (resultant and driver in phase). Since $\tan \phi$ is positive for $p < \omega$, ϕ must in this case lie between 0 and $\pi/2$. At resonance when $p = \omega$, $\tan \phi = \infty$ and $\phi = \pi/2$ (resultant lags in phase by 90° with respect to the driver).

The energy of the system undergoing forced vibrations
We have seen that for a system undergoing forced vibrations, the displacement x is given by $x = A \cos (pt - \phi)$, thus the velocity at any instant

$$\frac{dx}{dt} = -Ap \sin (pt - \phi)$$

The maximum velocity, or velocity amplitude is thus $-Ap$. Irrespective of damping, this maximum velocity occurs for zero displacement, and the corresponding kinetic energy is $\frac{1}{2}mA^2p^2$, where m is the mass of the system. Since the oscillations of the system are simple harmonic in the steady state, $\frac{1}{2}mA^2p^2$ is also the total energy E of the motion.

Now the velocity amplitude $v_0 = -AP$, is a function of p, and has a maximum obtained by differentiating with respect to p. Thus

$$\frac{dv_0}{dp} = \frac{\frac{1}{2}ap(4p^3 - 4\omega^2p + 8k^2p) - a[(\omega^2 - p^2)^2 + 4k^2p^2]}{[(\omega^2 - p^2)^2 + 4k^2p^2]^{3/2}} = 0$$

for a maximum. That is

$$\tfrac{1}{2}ap(4p^3 - 4\omega^2p + 8k^2p) = a(\omega^4 + p^4 - 2\omega^2p^2 + 4k^2p^2)$$

or

$$a(2p^4 - 2\omega^2p^2 + 4k^2p^2) = a(\omega^4 + p^4 - 2\omega^2p^2 + 4k^2p^2)$$

i.e.

$$ap^4 = a\omega^4$$

and $p = \omega$

Thus the total energy of the system is a maximum when $p = \omega$. Figure 7.14 shows how the total energy E varies with p, for dif-

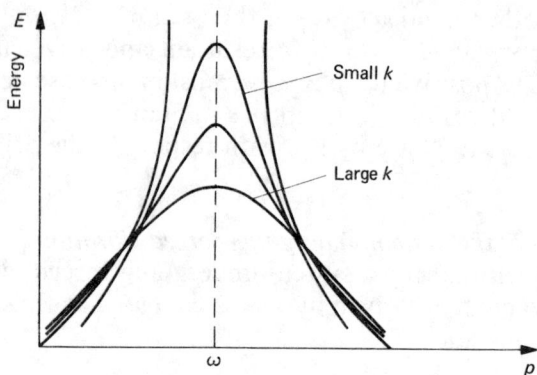

Fig. 7.14

ferent values of k. Notice that in contrast to the displacement
amplitude versus p curves, E approaches zero as $p \to 0$ irrespective of k.

Part 2   Heat

Part 2   Heat

8 Temperature

To measure the temperature of a body a scale of temperature is required, and some property of a substance which depends on temperature (thermometric property), e.g. the resistance of a wire, the volume of a gas, a thermal e.m.f. In addition the establishment of a temperature scale requires the use of 'fixed points'. These are constant and easily reproduced temperatures, e.g. the melting point of ice (ice point), the boiling point of water (steam point). For both the ice point and the steam point the pressure must be specified. This has been taken as 76 cm of mercury at 45° latitude (pressure depends on g which in turn depends on latitude).

8.1 A centigrade scale

The fundamental fixed points used in the case of a centigrade scale are the ice point (0 °C), and the steam point (100 °C). Let the thermometric property be X, its value at the ice point being X_0, and that at the steam point being X_{100}. Then $X_{100} - X_0$ is called the fundamental interval for the particular thermometer we are considering, and a unit interval is taken as $(X_{100} - X_0)/100$. If an environment of unknown temperature t_X, gives a value X_t for the thermometric property, then t_X is the number of unit intervals in $X_t - X_0$, i.e.

$$t_X = \frac{X_t - X_0}{(X_{100} - X_0)/100}$$

or $\quad t_X = 100 \, \dfrac{X_t - X_0}{X_{100} - X_0}$

This gives the temperature as measured by the particular thermometer we are considering, which will not in general be the same as the temperature measured by some other thermometer.

8.2 The condition for two thermometer scales to agree

We have seen that for a given thermometer $t_X = 100(X_t - X_0)/(X_{100} - X_0)$. Suppose now that we have another thermometer with thermometric property Y. Then with this thermometer, the environment which gave a temperature t_X with the first thermometer, now gives a temperature

$$t_Y = 100 \frac{Y_t - Y_0}{Y_{100} - Y_0}$$

If $t_x = t_y$, then

$$t_x = 100 \frac{Y - Y_0}{Y_{100} - Y_0} \quad \text{or} \quad t_y = 100 \frac{X - X_0}{X_{100} - X_0}$$

and

$$Y = Y_0 \left(1 + \frac{(Y_{100} - Y_0)}{100 Y_0} t_x \right)$$

This is a linear relation, Y being a linear function of t_x (Fig. 8.1).

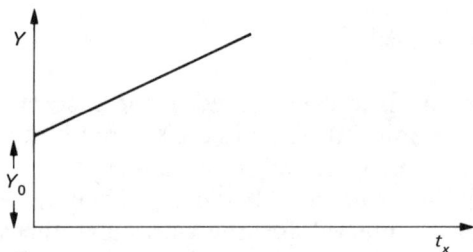

Fig. 8.1

8.3 Gas scales

The thermometer adopted as a universal standard must have high sensitivity. In fact, the pressure exerted by a gas, with the volume maintained constant, is very sensitive to changes in temperature. Alternatively the pressure can be maintained constant, and the volume used as the thermometric property. Thus a gas thermometer is used as a standard for accurate work.

At constant pressure (volume measured)

$$t_p = 100 \, \frac{V_t - V_0}{V_{100} - V_0}$$

where V denotes volume, while at constant volume (pressure measured, the preferred method in practice at the National Physical Laboratory)

$$t_v = 100 \, \frac{P_t - P_0}{P_{100} - P_0}$$

Now according to Boyle's law $PV = mk$ for a given temperature, where k is a constant depending on the kind of gas, and m is the mass of the gas. Thus to obtain Boyle products we multiply the top and bottom terms in our expression for t_p by the particular constant pressure P_c, i.e.

$$t_p = 100 \, \frac{P_c V_t - P_c V_0}{P_c V_{100} - P_c V_0}$$

By similar means, this time using the particular constant volume V_c used in the case of our constant-volume thermometer, we have

$$t_V = 100 \, \frac{P_t V_c - P_0 V_c}{P_{100} V_c - P_0 V_c}$$

Thus $t_P = t_V$ for a given gas if Boyle's law holds, since

$$t_P = 100 \, \frac{(m_P k_t - m_P k_0)}{(m_P k_{100} - m_P k_0)}$$

where m_P is the mass of gas used in the constant-pressure case, and

$$t_V = 100 \, \frac{m_V k_t - m_V k_0}{m_V k_{100} - m_V k_0}$$

where m_V is the mass of gas used in the constant-volume case. In fact a gas obeys Boyle's law more and more closely as the pressure is reduced to lower and lower values, a gas at indefinitely low pressure being an 'ideal' gas.

Thus for a given gas 1, as $P \to 0$, $t_P = t_V = t_1$. But it is found experimentally that

$$(PV)_1 = \left[(PV)_0 \right]_1 (1 + \alpha t_2) \quad P \to 0$$

where t_2 denotes temperature as measured by a gas thermometer using another gas 2, under conditions where $P \to 0$, where $[(PV)_0]_1$ is the Boyle product for gas 1 at the ice point, and where α is a constant independent of temperature (the fact that α is a constant the same for all gases is Charles' law). This means that the Boyle product as $P \to 0$ for one kind of gas, is a linear function of temperature as measured by another kind of gas, and from what we have said about the condition for two thermometer scales to agree, it follows that $t_1 = t_2$, where the thermometric property is PV. Temperature on the ideal gas scale is therefore defined as

$$t = 100 \underset{P \to 0}{} \frac{(PV)_t - (PV)_0}{(PV)_{100} - (PV)_0}$$

Either P or V is kept constant in practice when using a gas thermometer. We conclude then that a gas used under conditions where $P \to 0$ (ideal gas), gives the same temperature irrespective of the kind of gas (nitrogen, hydrogen etc.) used, or whether it is used in a constant-pressure or a constant-volume thermometer.

8.4   The thermodynamic or Kelvin temperature scale

It will be shown, when we deal with thermodynamics, that for a Carnot engine (ideal heat engine), if Q_1 is the heat absorbed from the hot reservoir, and Q_2 is the heat rejected to the cold reservoir, then Q_1/Q_2 is independent of the working substance used in the engine. The ratio of the temperatures of these reservoirs, however, depends on the particular thermometer used to make the measurement. Kelvin therefore proposed that the Carnot engine itself be used as a thermometer, and defined the thermodynamic temperatures of the hot and cold reservoirs (T_1, T_2), as $T_1/T_2 = Q_1/Q_2$. This thermodynamic scale of temperature has been adopted as the basic international scale, and to complete its definition, the arbitrary value 273·16K (K for Kelvin) has been assigned to the triple point of water T_3 (the temperature at which water, water vapour and ice co-exist in equilibrium), and absolute zero has been defined as 0K. Thus

$$\frac{Q}{Q_3} = \frac{T}{T_3}$$

and

$$T = 273 \cdot 16 \frac{Q}{Q_3} \text{ K}$$

where Q is the heat absorbed from the hot reservoir at temperature T, and Q_3 is the heat rejected to the cold reservoir at temperature T_3K. On this scale the ice point is $273 \cdot 16$K, and the
steam point is $373 \cdot 16$K. Thus the degree interval is identical
with the centigrade degree interval.

Now we will also show, when dealing with thermodynamics,
that the ratio of the two Kelvin temperatures

$$\frac{T_1}{T_2} = \frac{\theta_1}{\theta_2}$$

where θ_1 and θ_2 are ideal gas temperatures expressed in degrees
absolute (i.e. $273 \cdot 16$ is added to the centigrade temperature).
The centigrade ideal gas scale is known as the Celsius scale.
Thus, with the assumption that the triple point temperature
θ_3 is $273 \cdot 16°$ absolute, the ideal gas scale expressed in
degrees absolute is identical to the Kelvin scale, and

$$\theta = 273 \cdot 16 \frac{\lim_{P \to 0}(PV)}{\lim_{P \to 0}(PV)_3}$$

Note that the international scale uses the triple point of water
as a fixed point, in preference to the ice and steam points,
although these are still in general use. Thus Kelvin temperatures,
which in principle require the use of an ideal heat engine, can be
realised in practice by using an ideal gas. A value for PV as
$P \to 0$ at a given temperature is obtained by plotting PV against
P (Fig. 8.2) and extrapolating to obtain the zero pressure value
of PV.

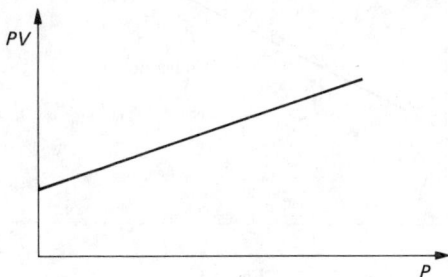

Fig. 8.2

8.5   Correction of a real gas temperature to the ideal gas temperature

A gas must be at an indefinitely low pressure for Boyle's law to be obeyed and since in practice it is not possible to use a gas thermometer in a nearly evacuated state, a correction is required to a real gas temperature to get the ideal gas temperature t. Now the Boyle product at a particular temperature is slightly higher than one would expect from Boyle's law, and increases as the pressure increases (Fig. 8.3).

In fact $(PV)_t = A_t + B_t P$ (the suffix t indicates the temperature concerned), and the straight line continues as the pressure is reduced indefinitely. A_t is the value of the Boyle product for zero pressure, and B_t is called the second virial coefficient. In general

$$(PV)_t = A_t + B_t P + C_t P^2 + D_t P^3$$

but we are well away from this situation in gas thermometry and only the first two terms are required (pressure sufficiently low that the C_t and D_t terms can be ignored).

Consider the constant pressure usage,

$$t_P = 100 \frac{V - V_0}{V_{100} - V_0} \quad \text{at constant pressure } P_c$$

or

$$t_P = 100 \frac{P_c V - P_c V_0}{P_c V_{100} - P_c V_0}$$

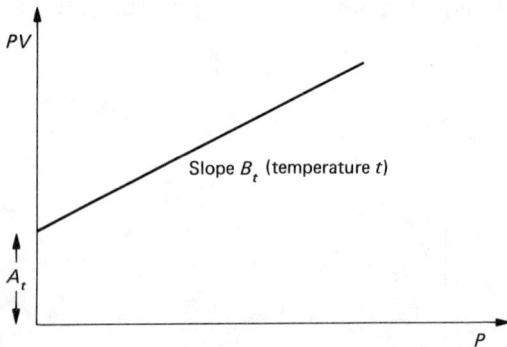

Fig. 8.3

But the ideal gas temperature

$$t = 100 \frac{A - A_0}{A_{100} - A_0}$$

where the A's are intercepts obtained from graphs such as that shown in Fig. 8.3. We now use the fact that $P_c V = A + BP_c$, giving

$$t_P = 100 \frac{(A + BP_c) - (A_0 + B_0 P_c)}{(A_{100} + B_{100} P_c) - (A_0 + B_0 P_c)}$$

The A and B without suffixes are for the unknown temperature. Therefore

$$t_P = 100 \frac{(A - A_0) + P_c(B - B_0)}{(A_{100} - A_0) + P_c(B_{100} - B_0)}$$

or

$$t_P = t \left[\frac{1 + [P_c(B - B_0)/(A - A_0)]}{1 + [P_c(B_{100} - B_0)/(A_{100} - A_0)]} \right]$$

where the A's and B's are obtained from the intercepts and slopes of graphs such as that shown in Fig. 8.3. Writing $B - B_0$ as ΔB and $A - A_0$ as ΔA we now assume $\Delta B/\Delta A \ll 1$. We also denote

$$a = P_c(B - B_0)/(A - A_0)$$

and

$$b = P_c(B_{100} - B_0)/(A_{100} - A_0)$$

We then have

$$t_P = t \left(\frac{1 + a}{1 + b} \right)$$

If $b \ll 1$ then $(1 + b)^{-1} \cong 1 - b$ by the binomial theorem, ignoring terms of order higher than 1. Thus

$$t_P \cong t[(1 + a)(1 - b)] \cong t(1 + a - b)$$

neglecting the small term ab, and substituting back for a and b we have

$$t_P - t = P_c t \left(\frac{B - B_0}{A - A_0} - \frac{B_{100} - B_0}{A_{100} - A_0} \right)$$

where the A's and B's are determined by experiment. Accurate temperature measurement in industry is not done like this, but is usually carried out by calibrating a platinum resistance thermometer using the sulphur point 444·6 °C.

8.6 The platinum resistance thermometer

It has long been established that pure platinum carefully annealed (under no strain), has a resistance R as a function of temperature, obeying the law

$$R = R_0(1 + at + bt^2)$$

where R_0 is the resistance at the ice point, a and b are constants, and t is the centigrade ideal gas temperature. But t_{Pt}, the temperature on the platinum centigrade scale is given by

$$t_{Pt} = 100 \frac{R - R_0}{R_{100} - R_0}$$

$$= \frac{100R_0(at + bt^2)}{R_0[a(100) + b(100)^2]}$$

$$= \frac{at + bt^2}{a + 100b}$$

Therefore

$$t_{Pt} - t = t\left[\frac{a + bt}{a + 100b} - 1\right]$$

$$= t\,\frac{b(t - 100)}{a + 100b}$$

$$= \frac{b}{a + 100b}(t^2 - 100t)$$

$$= \frac{10^4 b}{a + 100b}\left[\left(\frac{t}{100}\right)^2 - \frac{t}{100}\right]$$

$$= \delta\left[\left(\frac{t}{100}\right)^2 - \frac{t}{100}\right]$$

where δ is the calibration constant for the particular thermometer we are considering.

Manufacturers usually supply a table, giving values of t_{Pt} and the corresponding values of t. Failing this δ can be determined by measuring t_{Pt} at the sulphur point ($t = 444\cdot6$ °C), and then t, corresponding to the value of t_{Pt} at some other temperature, can be obtained by a method of successive approximations. The measured value t_{Pt} is put into the right-hand side of the equation for t, and a value for t, say t_1, is calculated. This approximate value of t, t_1, is now put into the right-hand side of the equation, and a more accurate value of t is calculated. This process is repeated until a sufficiently accurate value of t is obtained. Thus a convenient thermometer like the platinum resistance thermometer, can be used to obtain ideal gas temperatures, without the need to take the thermometer to the NPL for calibration. The resistance R of the platinum coil of the resistance thermometer, is measured by using a Callendar and Griffiths bridge (Fig. 8.4). The resistance r, in series with R, is the resistance of the leads connecting the coil to the bridge. The resistance r in series with R_3, is the resistance of a pair of dummy leads identical in every respect to the actual leads. Connected in this way, these dummy leads balance the resistance of the actual leads at all temperatures, and thus the effect of variation of resistance of connecting leads with temperature is eliminated.

To measure R, first BD and CE are shorted out, and R_1 set equal to R_2. This gives a balance point at the middle of the bridge wire. With BD and CE back in circuit again, say a balance point is obtained at a distance x to the left of the mid-

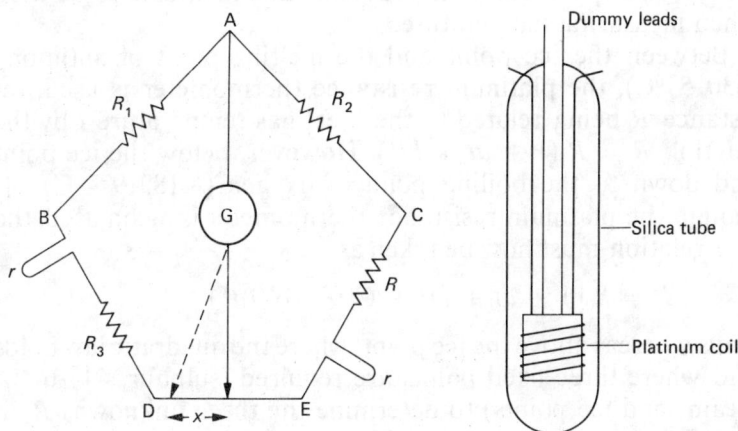

Fig. 8.4

point. Then for balance

$$\frac{R_1}{R_3 + r + \rho(l - x)} = \frac{R_2}{R + r + \rho(l + x)}$$

where l is the half-length of the bridge wire, and ρ is the resistance per unit length of wire. But $R_1 = R_2$, thus

$$R_3 + \rho(l - x) = R + \rho(l + x)$$

or

$$R_3 + \rho l - \rho x = R + \rho l + \rho x$$

i.e.

$$R = R_3 - 2\rho x$$

The platinum resistance thermometer is very accurate and covers a wide range of temperature ($-200\ °C$ to $1000\ °C$). It is easy to use but has the disadvantage of being bulky, and therefore of not being suitable for measuring temperature in a small region.

8.7   The international temperature scale (1948)

An international committee in 1948, defined an international temperature scale by specifying certain fixed points. The committee also specified the instruments and formulae to be used when measuring temperature.

Between the ice point and the melting point of antimony ($630.5\ °C$), the platinum resistance thermometer is used, resistance R being related to the ideal gas temperature t by the relation $R = R_0(1 + at + bt^2)$. However, below the ice point and down to the boiling point of oxygen ($-182.97\ °C$), although the platinum resistance thermometer is again used, the R, t relation must now be taken as

$$R = R_0\left[1 + at + bt^2 + C(t - 100)t^3\right]$$

Thus whereas above the ice point, where the quadratic law holds and where three fixed points are required (sulphur, $444.6\ °C$, steam, and ice points) to determine the three unknowns R_0, a and b, below the ice point a further fixed point is required to determine the extra unknown C. The extra fixed point used is the boiling point of oxygen. There is not sufficient confidence in measuring temperatures below −183 °C to define these on an international scale.

Above the antimony point, and up to the melting point of gold (1063 °C), the platinum–platinum rhodium thermocouple is used, the e.m.f. e obeying the law $e = a + bt + ct^2$, where t is the ideal gas temperature. Three fixed points are required to determine the constants a, b and c: the antimony point, the gold point and the melting point of silver (960·8 °C).

Above the melting point of gold, a furnace is used and Planck's radiation law (to be discussed when we deal with radiation). For these very high temperatures no container could hold a gas without softening, and thus a gas thermometer cannot be used. The international scale is now defined for indefinitely high temperatures, since the measurement of very high temperatures is required for example in the case of a hydrogen bomb.

8.8 Environments used for various temperature ranges

In the region −180 °C to −60 °C the environment used is dry air. From −60 °C to room temperature (low temperatures used in chemistry) a mixture of solid carbon dioxide and acetone is used.

From room temperature to 70 °C water is suitable, and from 70 °C to 240 °C paraffin oil, or some other oil is used. From 240 °C to 2000 °C the environment is molten tin, and above 2000 °C a furnace with hot air is used.

9 The Kinetic Theory of Gases

We shall consider an ideal gas, in which case the simplifying assumptions to be made are as follows:

(1) The volume of the molecules is negligible compared with the volume of the gas, i.e. the main body of the gas is empty space.
(2) Attractive forces between the molecules are negligible. Molecules only exert forces on one another on collision. Therefore the molecules move in straight lines.
(3) The molecules behave like perfectly elastic spheres, i.e. the total energy after a collision equals the total energy before the collision. Momentum is also conserved.
(4) The time of collision is negligible compared with the time between collisions.
(5) All molecules of the same gas have the same mass.
(6) After collisions, molecules have random directions and speeds.
(7) Steady state conditions obtain with a uniform density.

A real gas will approximate more and more closely to an ideal gas as the pressure tends to zero, since the lower the pressure, the more negligible are the attractive forces between molecules. This is not the only reason for a gas becoming more nearly perfect as the pressure is lowered: the factors (1), (2) and (4) are all involved.

Let the mass of a molecule be m', the number of molecules per unit volume be n, the total number being N, and let M be the kilogram molecular weight of the gas.

9.1 Calculation of the pressure exerted by an ideal gas

Since force equals rate of change of momentum, in order to obtain the pressure we must calculate the rate of change of momentum per unit area. But this will depend on the velocity of the molecules, so we will consider the group of molecules with the same velocity C_1 (n_1 per unit volume with this velocity).

Now if we consider an element of area, the molecules will approach this area from all directions (all values of θ) (Fig. 9.1).

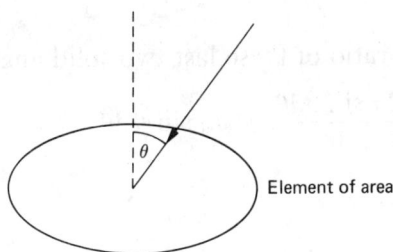

Element of area

Fig. 9.1

Thus we first require the number of molecules per unit volume with velocity C_1, which are approaching with directions between θ and $\theta + d\theta$. To obtain this we must consider the concept of solid angle.

The solid angle subtended by the element of area da, at the point O, is $da \cos \theta / R^2$ (by definition) (Fig. 9.2). Thus a sphere will subtend a solid angle $4\pi R^2 / R^2 = 4\pi$ (R is the radius of the sphere) at the centre.

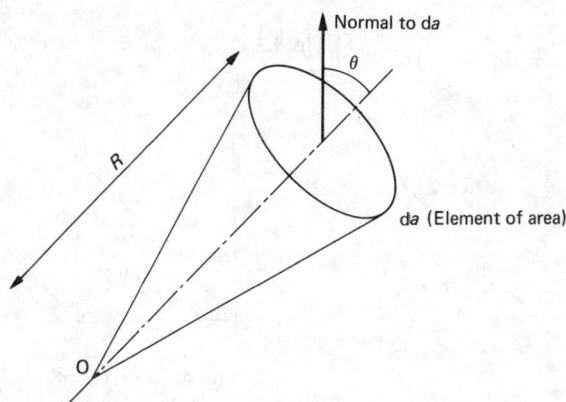

Normal to da

θ

R

da (Element of area)

O

Fig. 9.2

The solid angle subtended at the centre of the sphere, by the shaded strip defined by angle $\theta \to \theta + d\theta$, is given by

$$\frac{2\pi R \sin \theta R d\theta}{R^2} = 2\pi \sin \theta d\theta$$

(Fig. 9.3). Therefore the number of molecules per unit volume with velocity C_1, and with directions $\theta \to \theta + d\theta$ (i.e. C_1, θ molecules) is

$n_1 \times$ the ratio of these last two solid angles

$$= n_1 \times \frac{2\pi \sin \theta d\theta}{4\pi} = \tfrac{1}{2}n_1 \sin \theta \, d\theta$$

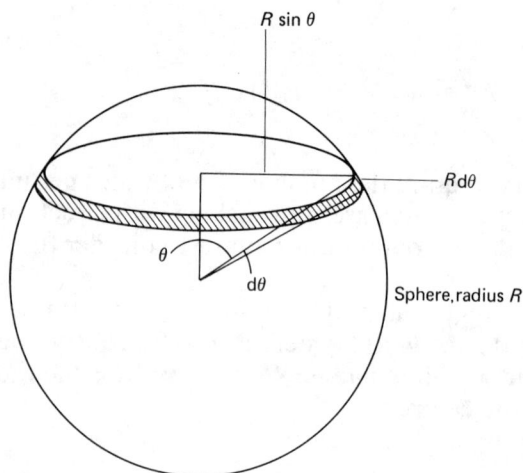

Fig. 9.3

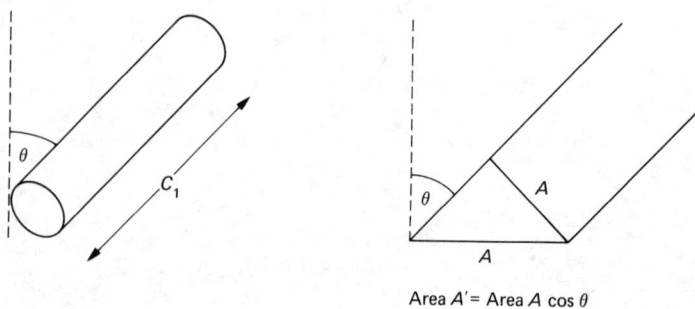

Area $A' =$ Area $A \cos \theta$

Fig. 9.4

With reference to Fig. 9.4, the number of C_1, 'θ' molecules striking unit area per second is $\frac{1}{2}n_1 \sin \theta \, d\theta (C_1 \cos \theta)$. For each of these molecules the normal component of momentum is $m'C_1 \cos \theta$, and the change of momentum following a collision is $2m'C_1 \cos \theta$ (Fig. 9.5). Thus the change of momentum per second per unit area for the 'θ' molecules, with velocity C_1 is

$$dP_1 = \frac{1}{2}n_1 \sin \theta \, d\theta \times C_1 \cos \theta \times 2m'C_1 \cos \theta$$
$$= m'n_1 C_1^2 \cos^2 \theta \sin \theta \, d\theta$$

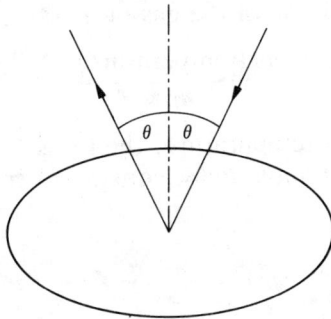

Fig. 9.5

Therefore for molecules with velocity C_1, but considering all directions of approach, the pressure is given by integrating over all values that θ takes:

$$P_1 = - \int_0^{\pi/2} m'n_1 C_1^2 \cos^2 \theta \, d(\cos \theta)$$

replacing $\sin \theta \, d\theta$ by $-d(\cos \theta)$. Thus $P_1 = \frac{1}{3}m'n_1 C_1^2$. Finally the total pressure P, considering all velocities, is given by

$$P = \frac{1}{3}m' \Sigma n_1 C_1^2$$

But the mean square velocity is defined as

$$\overline{C^2} = \frac{\Sigma n_1 C_1^2}{n}$$

Therefore

$$P = \frac{1}{3}m'n\overline{C^2}$$

9.2   Deduction of the gas laws from kinetic theory

Boyle's law

$P = \frac{1}{3}\rho\,\overline{C^2}$, since nm' is the density of the gas ρ.

We now make the assumption that $m'\overline{C^2}$, which depends on the average kinetic energy of the molecules, is some function of temperature f(temperature). Therefore

$$\overline{C^2} = \frac{f\,(\text{temperature})}{m'}$$

Also, if v is the volume of the gas of mass m, we can write

$$Pv = \frac{1}{3}m\overline{C^2} = \frac{\frac{1}{3}mf(\text{temperature})}{m'}$$

Thus if we fix the temperature, the mass of gas and the kind of gas (i.e. m'), we have $Pv = $ constant, which is Boyle's law.

Avogadro's law
We have seen that

$$P = \frac{1}{3}nm'\,\overline{C^2}$$

Therefore $Pv = \frac{1}{3}m'N\overline{C^2}$, where N is the number of molecules in the volume v. But $m'\overline{C^2} = f$(temperature). Therefore $Pv = Nf$(temperature), and it follows that equal volumes of all gases, under the same conditions of temperature and pressure, contain the same number of molecules N. This is Avogadro's law.

Now we have

$$P = \frac{1}{3}\rho\,\overline{C^2} = \frac{1}{3}(m/v)\,\overline{C^2}$$

where m is the mass of the gas and v its volume. Thus for one kilogram mole we may write $PV = \frac{1}{3}M\overline{C^2} = RT$, where M is the kilogram molecular weight of the gas, V is its volume, R is the universal gas constant and T is the absolute temperature. Therefore

$$\overline{C^2} = 3RT/\text{M} \propto T$$

but independent of P and V, and

$$\overline{C^2} = 3\,\frac{RT}{M} = \frac{3R}{N_0m'}\,T$$

where N_0 is Avogadro's number, the number of molecules in one kilogram mole of the gas, and m' is the mass of one molecule. Thus

$$\tfrac{1}{3} m' \, \overline{C^2} = \frac{R}{N_0} T = kT,$$

where k is Boltzmann's constant, and

$$\tfrac{1}{2} m' \, \overline{C^2} = \tfrac{3}{2} kT.$$

Graham's Law

We have $\overline{C^2} = 3P/\rho$, where ρ is the density of the gas. Therefore the root mean square velocity r.m.s. velocity

$$= \sqrt{\frac{3P}{\rho}}$$

Now a gas flows through a fine hole at a rate depending on its mean velocity $\propto$ r.m.s. velocity. Therefore

$$\text{rate of effusion} \propto \frac{1}{\sqrt{\text{density}}} \text{ (Graham's law)}.$$

Dalton's Law of Partial Pressures

The partial pressure of a gas in a gas mixture, is the pressure the gas would exert of it occupied the volume of the container on its own.

Suppose we have a mixture of gases in a volume V, exerting pressures $P_1, P_2, \ldots$ respectively.

Then $P_1 V = \tfrac{1}{3} N_1 \, m_1' \, \overline{C_1^2}$, where N_1 is the number of molecules of gas 1, and m_1' is the mass of a molecule of gas 1. Similarly $P_2 V = \tfrac{1}{3} N_2 \, m_2' \, \overline{C_2^2}$ etc.

Now the average kinetic energy of the different gases is the same since the temperature is the same, i.e. $m_1' \overline{C_1^2} = m_2' \overline{C_2^2} = m \, \overline{C^2}$ say.

Therefore

$$(P_1 + P_2 + \cdots) V = \tfrac{1}{3}(N_1 + N_2 + \cdots) m \overline{C^2}$$
$$= \tfrac{1}{3} N m \overline{C^2},$$

where N is the total number of molecules. Therefore the total pressure of the gas is $P_1 + P_2 + \cdots$ which is the sum of the partial pressures.

9.3 The kinetic theory derivation of the thermal conductivity of a gas

A temperature gradient is required in order to have a flow of heat. Consider unit area in a plane whose x coordinate is x_0 (Fig. 9.6). E is the average energy of a molecule in this plane. As in

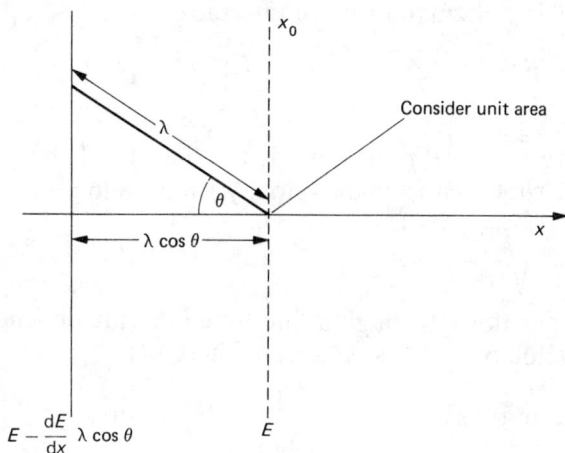

Fig. 9.6

the case of the pressure derivation, the number of molecules per unit volume with velocity C_1 approaching our unit area with directions between θ and $\theta + d\theta$, (i.e. C_1, 'θ' molecules) is $\frac{1}{2}n_1 \sin \theta d\theta$ (Fig. 9.7), where n_1 is the number of molecules per unit volume with velocity C_1. With reference to Fig. 9.8 the number of C_1, 'θ' molecules incident on our unit area per second from one side is given by

$$\tfrac{1}{2}n_1 \sin \theta d\theta C_1 \cos \theta = \tfrac{1}{2}n_1 C_1 \cos \theta \sin \theta d\theta$$

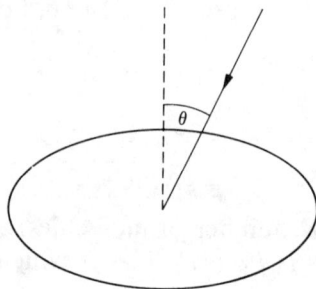

Fig. 9.7

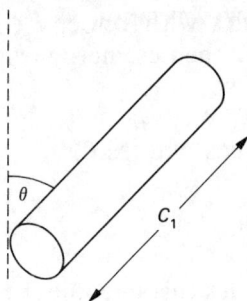

Fig. 9.8

These molecules bring with them energy, and we assume each molecule crosses the plane with x coordinate x_0, with the energy it possesses from its last collision. If λ is the mean free path (the average distance between collisions), we consider molecules which on the average come from planes whose x coordinates are $x_0 \pm \lambda \cos \theta$ (Fig. 9.9). Thus the net energy transferred per unit area per second from left to right by the C_1, 'θ' molecules is

$$\tfrac{1}{2} n_1 C_1 \cos \theta \sin \theta \mathrm{d}\theta \left(-2 \frac{\mathrm{d}E}{\mathrm{d}x} \lambda \cos \theta \right)$$

$$= -n_1 C_1 \frac{\mathrm{d}E}{\mathrm{d}x} \lambda \cos^2 \theta \sin \theta \mathrm{d}\theta$$

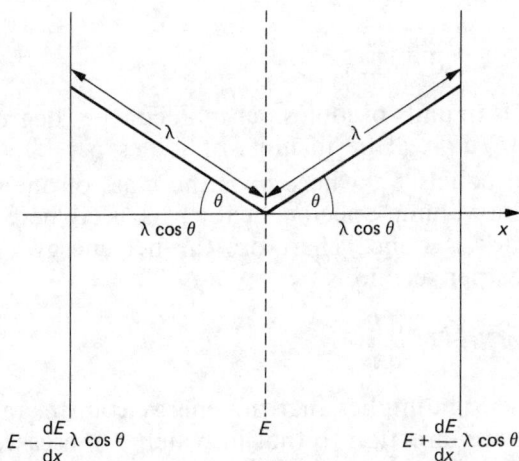

Fig. 9.9

Therefore for molecules with velocity C_1, but considering all directions of approach, the net energy transferred per unit area per second is

$$\int_0^{\pi/2} n_1 C_1 \frac{dE}{dx} \lambda \cos^2 \theta d(\cos \theta)$$

$$= -\tfrac{1}{3} n_1 C_1 \frac{dE}{dx} \lambda$$

Considering all velocities this becomes

$$-\frac{1}{3} \frac{dE}{dx} \lambda \Sigma n_1 C_1$$

$$= -\tfrac{1}{3} n \overline{C} \lambda \frac{dE}{dx}$$

since the mean velocity $\overline{C}$ is

$$\overline{C} = \frac{\Sigma n_1 C_1}{n},$$

where n is the total number of molecules per unit volume.

Now in order to introduce the temperature gradient, we note that

$$\frac{dE}{dx} = \frac{dE}{dT} \frac{dT}{dx}$$

where T denotes temperature. Thus the net energy transferred per unit area per second is

$$-\frac{1}{3} n \overline{C} \lambda \frac{dE}{dT} \frac{dT}{dx}$$

But dE/dT is in units of joules per molecule per degree. Therefor $(1/m') dE/dT$ will be in units of joules per kilogram, per degree, and equals C_v, where m' is the mass of one molecule. The constant-volume specific heat C_v is used here since no external work is done. Therefore the net energy transferred per unit area per second is

$$-\frac{1}{3} n \overline{C} \lambda m' C_v \frac{dT}{dx}$$

The negative sign implies that the energy transfer takes place in the opposite direction to that in which T increases. But the energy transferred per second per unit area is given by $-k\,dT/dx$, where k is the thermal conductivity of the gas. Therefore the thermal conductivity of the gas is given by

$$k = \tfrac{1}{3}n\overline{C}\lambda m'C_v$$

9.4 Mean free path λ

As a first approximation, we imagine that a molecule is a sphere of diameter S. Assume that a particular molecule moves through the gas and has collisions with other molecules which are at rest. Relative velocity is the important quantity here.

With reference to Fig. 9.10 our molecule will collide with all molecules whose centres lie inside a cylinder of radius S. Therefore the number of collisions per unit distance is $n\,(\pi S^2 \times 1)$, where n is the number of molecules per unit volume. Thus the distance between collisions $\lambda = 1/n\pi S^2$. The fact that all molecules are in motion enhances the number of collisions made in a given time. Also account should be taken of the distribution of velocities. In fact a better approximation gives the mean free path as $\lambda = 1/\sqrt{2}n\pi S^2$. Since λ is proportional to $1/n$, for a given gas it will be inversely proportional to the density and therefore inversely proportional to the gas pressure as long as Boyle's law applies. Note that $n\lambda \propto 1/\pi S^2$, and is a constant for a given molecule.

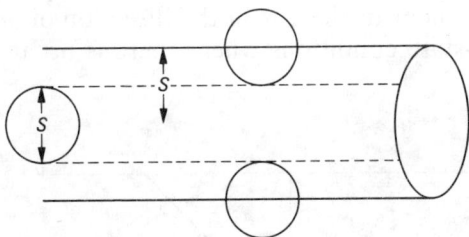

Fig. 9.10

Looking at our expression for the thermal conductivity of a gas, we see that this is proportional to $\overline{C}$, m' and C_v, but is independent of the density, and therefore of the pressure of the gas, since $n\lambda$ is a constant. In fact since $\overline{C}$ is proportional to $\sqrt{T}$, as we shall see later, where T denotes absolute temperature, the thermal conductivity is proportional to $\sqrt{T}$.

These deductions are correct in as much as S is not a function of pressure or temperature. In fact statistically S is a function of both pressure and temperature. If the molecules are moving rapidly, this has the effect of reducing the apparent diameter S. Also at high pressures, where the gas has a higher density and where the molecules are in close proximity, molecular attractive forces come into play. Our theory is the ideal gas theory where the assumption is made that there are no attractive forces between molecules. Thus our prediction that the thermal conductivity is independent of pressure is wrong at high pressures, since attractive forces are present. It is also wrong at low pressures, where the thermal conductivity is found to be proportional to pressure. Also experimentally it is found that, depending on the gas, some modification has to be made to the factor $\frac{1}{3}$ which appears in our expression for the thermal conductivity.

9.5   The kinetic theory derivation of the viscosity of a gas

Considering the viscous flow of gas, different layers of the gas have different drift velocities in the direction of flow. Let U be the drift velocity of the layer where z coordinate is z_0 (Fig. 9.11). This drift velocity is superimposed on the random thermal velocity of the molecules and is responsible for the general movement of the gas in the direction of flow. We consider steady-state conditions where there is no acceleration of the gas.

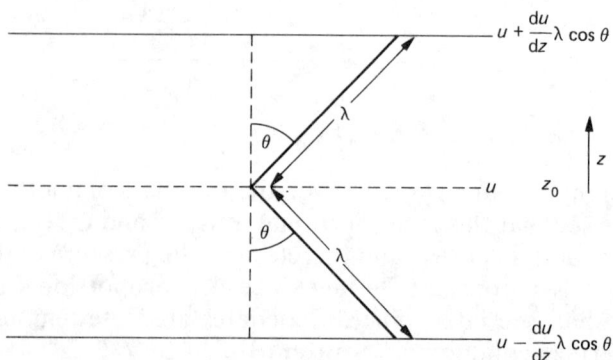

$$u + \frac{du}{dz}\lambda \cos\theta$$

$$u$$

$$u - \frac{du}{dz}\lambda \cos\theta$$

Fig. 9.11

As in the case of the pressure and thermal conductivity derivations, the number of molecules of the (C_1, θ) group entering the U layer per unit area per second, from one side, is $\frac{1}{2}n_1 \sin \theta \, d\theta \, C_1 \cos \theta$, where n_1 is the number of molecules per unit volume with a velocity C_1. If λ is the mean free path, these molecules on average come from a plane whose z coordinate is $z_0 + \lambda \cos \theta$. The same number of C_1, 'θ' molecules enter the U layer from the other side, and these on average come from a plane whose z coordinate is $z_0 - \lambda \cos \theta$. Each molecule carries with it from above drift momentum

$$m' \left(U + \frac{dU}{dz} \lambda \cos \theta \right)$$

and from below drift momentum

$$m' \left(U - \frac{dU}{dz} \lambda \cos \theta \right)$$

where m' is the mass of one molecule. Therefore the net supply of momentum to the U layer per unit area per second, from below to above by the C_1, 'θ' molecules, is

$$- \frac{1}{2}n_1 C_1 \sin \theta \cos \theta d\theta \left(2m' \frac{dU}{dz} \lambda \cos \theta \right)$$

Therefore for molecules with velocity C_1, but considering all directions of approach, the net drift momentum transferred to the U layer per unit area per second is

$$\int_0^{\pi/2} n_1 C_1 m' \lambda \frac{dU}{dz} \cos^2\theta d(\cos \theta) = -\frac{1}{3}n_1 C_1 m' \lambda \frac{dU}{dz}.$$

Considering all velocities this becomes

$$-\frac{1}{3}m' \lambda \frac{dU}{dZ} \Sigma n_1 C_1 = -\frac{1}{3}n\overline{C}m' \lambda \frac{dU}{dZ}$$

since the mean velocity $\overline{C}$ is $\overline{C} = \dfrac{\Sigma n_1 C_1}{n}$

where n is the total number of molecules per unit volume. The negative sign implies that the momentum transfer takes place in the opposite direction to that in which U increases.

But force is equal to the rate of change of momentum, and tangential frictional force per unit area is given by $\eta \times$ velocity gradient, where η is the coefficient of viscosity of the gas. Therefore by comparison

$$\eta = \tfrac{1}{3}n\,\overline{C}\,m'\lambda$$

As discussed in the case of thermal conductivity, for a given gas, $n\lambda$ should only depend on the molecule, since $n\lambda = 1/\epsilon\pi S^2$, where ϵ is some number and S is the effective diameter of the molecule. This means that η should be independent of the density and therefore independent of the pressure of the gas. This is in fact the case over a wide range of pressures. Also since $\overline{C} \propto \sqrt{T}$, where T is the absolute temperature, our theory predicts $\eta \propto \sqrt{T}$. However, we have assumed that the effective molecular diameter S is constant, whereas in fact S is a function of both pressure and temperature.

At high pressures our prediction that η is independent of pressure is invalid because attractive forces come into play. The prediction is also wrong at very low pressures because the mean free path λ may then be the same order of magnitude as the dimensions of the apparatus, in which case our theory breaks down. Various workers, including Maxwell and Jeans, have derived expressions for η, by taking into account the distribution of velocities, which we have ignored. However, these expressions for η only differ from the one we have derived in the numerical factor $\tfrac{1}{3}$. The values for this numerical factor range from $\tfrac{1}{3}$ to $0\cdot499$. Measurement of viscosity, apart from the uncertainty in the factor $\tfrac{1}{3}$, is the best way of deducing the mean free path λ.

9.6   The specific heats of gases

The internal energy of a gas is given by

$$U = \sum \tfrac{1}{2}mv^2 + \sum \text{PE}$$

where the first term is the sum of the kinetic energies of the molecules, and the second term represents the potential energy of the molecules.

Heat ΔQ supplied to a gas can either increase U or do external work ΔW (Fig. 9.12), i.e.

$$\Delta Q = \Delta U + \Delta W$$

The specific heat is $\Delta Q/\Delta T$ where T denotes temperature, and if we specify constant-volume conditions, i.e. no external work is

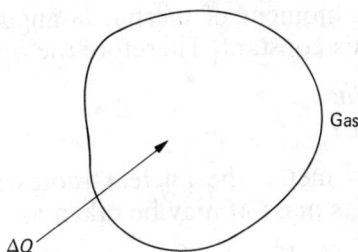

Fig. 9.12

done, then the specific heat at constant volume

$$C_V = \left(\frac{\partial U}{\partial T}\right)_V$$

9.7 Degrees of freedom

The number of degrees of freedom of a system is the number of independent squared terms which enter into the expression for the total energy of the system.

If we consider a monatomic gas (molecules consist of single atoms), the kinetic energy of the molecule is $\frac{1}{2}mu^2 + \frac{1}{2}mv^2 + \frac{1}{2}mw^2$, where the component velocities u, v and w can be varied independently. Each molecule therefore has three degrees of freedom. For N molecules, there will be $3N$ degrees of freedom, and these degrees of freedom are assigned to the translational movement. In addition however, rotation may occur independently about three axes of rotation, in which case a molecule of gas would have kinetic energy

$$\frac{1}{2}mu^2 + \frac{1}{2}mv^2 + \frac{1}{2}mw^2 + \frac{1}{2}I_1\omega_1^2 + \frac{1}{2}I_2\omega_2^2 + \frac{1}{2}I_3\omega_3^2$$

where I denotes moment of inertia, and ω denotes angular velocity; In this case the molecule would have six degrees of freedom. However, in the case of a monatomic gas the molecules cannot be set into rotation and hence there are no rotational degrees of freedom to consider. The reason for this is as follows.

According to the Bohr theory, the angular momentum of a rotating body is quantized according to the law

$$I\omega = n\,h/2\pi \quad n = 1, 2, 3 \ldots$$

where I denotes moment of inertia, ω angular velocity, and where h is Planck's constant. Therefore the rotational energy is

$$\tfrac{1}{2}I\omega^2 = \frac{n^2h^2}{8\pi^2 I}$$

Now for a heavy metal, the nuclear radius is approximately 10^{-14} metre, and its mass M may be taken as

$$100 \times 1 \cdot 67 \times 10^{-27}\,\text{kg} = 1 \cdot 67 \times 10^{-25}\,\text{kg}$$

Thus, for a sphere of radius R, $I = \tfrac{2}{5}MR^2$ and in the case of a heavy metal we have

$$I = \tfrac{2}{5} \times 1 \cdot 67 \times 10^{-25} \times 10^{-28}$$

$$= 0 \cdot 67 \times 10^{-53}\,\text{kg m}^2$$

Therefore the minimum rotational energy, given by putting $n = 1$ is approximately

$$\frac{h^2}{8 \times 9 \times 0 \cdot 67 \times 10^{-53}} = \frac{(6 \cdot 6 \times 10^{-34})^2}{72 \times 0 \cdot 67 \times 10^{-53}} \cong 10^{-15}\,\text{J}$$

In the case of a gas molecule which will have a smaller moment of inertia, the minimum rotational energy will be greater than this.

However, the average translational energy of a gas molecule at temperature T is $\tfrac{3}{2}kT$, where k is Boltzmann's constant. This equals

$$\tfrac{3}{2} \times 1 \cdot 38 \times 10^{-23}\,T \cong 2 \times 10^{-23}\,T \quad J$$

We see then that the minimum rotational energy is very high, and compared with it the average translational energy of a gas molecule is very small, even at the highest temperatures. Thus in the case of a monatomic gas, there is not enough energy in the translational movement to excite rotation, and therefore no rotation occurs; thus a monatomic gas has only three degress of freedom.

Diatomic gases

Consider two atoms a fixed distance apart, as in the case of a diatomic molecule (Fig. 9.13). Such a molecule will have three degrees of freedom of translation as in the case of a monatomic molecule. However, in addition, the molecule can rotate about any axis at right angles to the line joining the two atoms, but

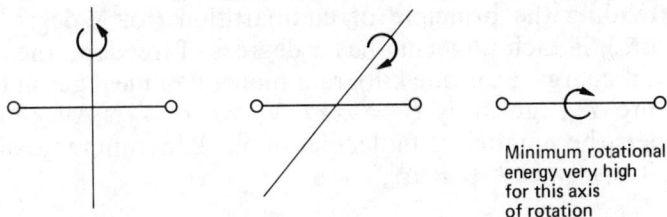

Minimum rotational
energy very high
for this axis
of rotation

Fig. 9.13

not about the line itself since the minimum rotational energy is too high for this last axis. Hence the diatomic molecule possesses five degrees of freedom, three of translation and two of rotation. N molecules have $5N$ degrees of freedom. However, the molecule may not always rotate, even about the two possible axes. In addition if the two atoms of a diatomic molecule can vibrate relative to each other, two extra degrees of freedom are added, one for kinetic energy and one for potential energy, since the energy of vibration is $\frac{1}{2}m\dot{x}^2 + \frac{1}{2}kx^2$, where k is a constant. Thus the maximum number of degrees of freedom for a diatomic molecule is seven.

Polyatomic gases
If the atoms of a polyatomic molecule are arranged in space at fixed distances apart, the molecule will have six degrees of freedom, three for translation and three for rotation. Also there will be two additional degrees of freedom for each pair of vibrating atoms.

9.8 The principle of equipartition of energy

Maxwell showed that if molecules obey the ordinary laws of mechanics, the total energy of a system is equally divided among the different degrees of freedom. Now for an ideal gas at absolute temperature T,

$$\tfrac{1}{2}m'\overline{C^2} = \tfrac{3}{2}kT$$

where k is Boltzmann's constant. Hence the translational energy on average is $\frac{3}{2}kT$. Thus by the principle of equipartition of energy, each degree of freedom contributes $\frac{1}{2}kT$ to the total energy.

Extending the principle of equipartition (for x degrees of freedom), if each molecule has x degrees of freedom, the total internal energy U of one kilogram mole of an ideal gas at temperature T, is given by $U = \frac{1}{2}xkT\,N_0$, where N_0 is Avogadro's number (the number of molecules in one kilogram mole of the gas). Thus, since $k = R/N_0$,

$$U = \frac{x}{2}\frac{R}{N_0}T\,N_0$$

$$= \frac{x}{2}RT$$

where R is the universal gas constant, and the molar specific heat at constant volume (the derivative of U at constant volume)

$$C_V = \left(\frac{\partial U}{\partial T}\right)_V$$

$$= \tfrac{1}{2}xR \text{ J mol}^{-1}\text{ K}^{-1}$$

But $C_p - C_V = R$, where C_p is the constant-pressure specific heat, (we shall derive this expression when we deal with thermodynamics). Therefore

$$C_V = \tfrac{1}{2}x(C_p - C_V)$$

Table 9.1　Values of C_V at 15 °C.

Gas	Number of atoms	C_V(kcal mol^{-1} K^{-1})	γ
Ar	1	2·98	1·667
He	1	2·98	1·667
H_2	2	4·87	1·405
HCl	2	5·11	1·400
N_2	2	4·93	1·401
O_2	2	5·04	1·396
CL_2	2	5·93	1·355
CO_2	3	6·75	1·300

Table 9.2   Theoretical values.

Degrees of freedom	C_V	$\gamma = 1 + (2/x)$
3	$\frac{3}{2}R = 2 \cdot 98$	$1 \cdot 667$
5	$\frac{5}{2}R = 4 \cdot 967$	$1 \cdot 400$
6	$3R = 5 \cdot 961$	$1 \cdot 333$
7	$\frac{7}{2}R = 6 \cdot 954$	$1 \cdot 286$

Now, if $\gamma = C_p/C_V$ we have

$$1 = \tfrac{1}{2}x(\gamma - 1)$$

that is

$$\gamma = 1 + \frac{2}{x}$$

Looking at Tables 9.1 and 9.2, there is good agreement between experimental and theoretical values for monatomic gases, assuming three degrees of freedom of translation. There is also quite good agreement for diatomic gases, assuming five degrees of freedom, three of translation, and two of rotation. At high temperatures C_V for diatomic gases may be approximately $\frac{7}{2}R$, since the translational energy may then be high enough to excite vibration, introducing two further degrees of freedom. For polyatomic gases there should be six degrees of freedom. Any possible oscillation adds two to the number of degrees of freedom. Polyatomic gases fit our theory only qualitatively.

For chlorine, which is diatomic, C_V is higher than expected. However, if we assume that there are vibrations giving seven degrees of freedom, then C_V is lower than expected. The experimental value corresponds to a fraction of a degree of freedom and a similar effect is shown by all the polyatomic gases. Classical mechanics is unable to explain the existence of these fractional degrees of freedom, but they can be explained by an application of quantum theory to the problem.

Effects at low temperatures
At low temperatures hydrogen behaves as a monatomic gas. This is explained by assuming that two degrees of freedom have

been lost, since the translational energy is not high enough at low temperatures to excite rotation. This loss of certain degrees of freedom is quite general at low temperatures.

9.9   The specific heats of solids

In most solids the atoms or molecules are arranged in a crystalline lattice, the molecules vibrating about their equilibrium positions. Since we are dealing with small displacements, we will consider that each atom performs simple harmonic motion. By the principle of equipartition of energy, the average kinetic energy of an atom is $\frac{3}{2}kT$ (motion is a superposition of vibrations in three mutually perpendicular directions). But the time average of the kinetic energy equals the time average of the potential energy. Therefore the average potential energy equals $\frac{3}{2}kT$. The total energy U per kilogram mole, is thus $\psi + 3RT$, where ψ is the potential energy of the atoms with respect to each other. ψ is a constant at constant volume, and thus

$$C_V = \left(\frac{\partial U}{\partial T}\right)_V$$

$$= 3R$$

$$= 5 \cdot 96 \text{ kcal mol}^{-1} \text{ K}^{-1}$$

This value agrees with the findings of Dulong and Petit (1818), who showed before any theory was developed, that for a large number of metals $C_V = 6$ kcal mol^{-1} K^{-1}.

9.10   The quantum theory of specific heat

At low temperatures deviations from classical theory occur: C_V being less than $3R$ and approaching zero at very low temperatures. According to Einstein (1905), who gave an explanation for the decrease in specific heat at low temperatures, the vibrations of the atoms in a solid must be treated quantum mechanically. In this case the average energy of a vibrator of frequency

f at temperature T, apart from the so called zero-point energy $\frac{1}{2}hf$, is

$$\frac{hf}{\exp(hf/kT) - 1}$$

But since each atom has three degrees of freedom the total energy of the thermal vibrations of one kilogram mole of a crystal, less the total zero-point energy, equals

$$U - U_0 = \frac{3N_0 hf}{\exp(hf/kT) - 1}$$

where N_0 is Avogadro's number. Thus

$$U - U_0 = 3RT \frac{\psi/T}{\exp(\psi/T) - 1}$$

where $\psi = hf/k$ is the characteristic temperature of the solid. At high temperatures $U - U_0 \to 3RT$. At low temperatures $U - U_0 \to 0$ and

$$C_V = \frac{3R(\psi/T)^2 e^{\psi/T}}{\left[\exp(\psi/T) - 1\right]^2}$$

With reference to Fig. 9.14, ψ is chosen so that the curve fits the experimental data at high temperatures. Einstein's theory agrees with experiment very well at high temperatures but there is a discrepancy at low temperatures. This discrepancy was removed by Debye in 1912. Debye ignored discrete atomic

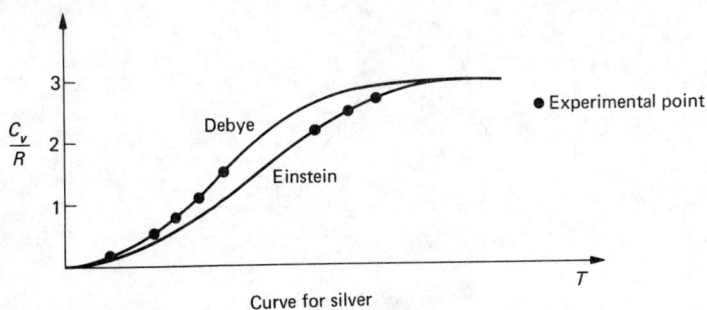

Curve for silver

Fig. 9.14

structure and treated a solid as a continuous elastic medium. He obtained good agreement with experiment even at very low temperatures.

9.11 The Maxwell–Boltzmann velocity distribution

In a gas different molecules have different velocities and the Maxwell–Boltzmann velocity distribution law is given by

$$\frac{dN}{N} = v^2 \left(\frac{2}{\pi}\right)^{1/2} \left(\frac{m}{kT}\right)^{3/2} \exp\left(\frac{-mv^2}{2kT}\right) dv$$

where v denotes velocity, T denotes absolute temperature, k is Boltzmann's constant, N is the total number of molecules and dN/N is the fraction of the molecules which have speeds between v and $v + dv$. Thus

$$dN = N\left(\frac{2m^3}{\pi k^3 T^3}\right)^{1/2} v^2 \exp\left(\frac{-mv^2}{2kT}\right) dv,$$

and the number of molecules with speeds between v and $v + dv$ is

$$Av^2 e^{-\beta v^2} dv = f(v)$$

where A is a constant and $\beta = m/2kT$.

Referring to Fig. 9.15, the shaded area represents the number of molecules with speeds between v and $v + dv$. The speed for which the curve is a maximum is the most probable speed v_m:

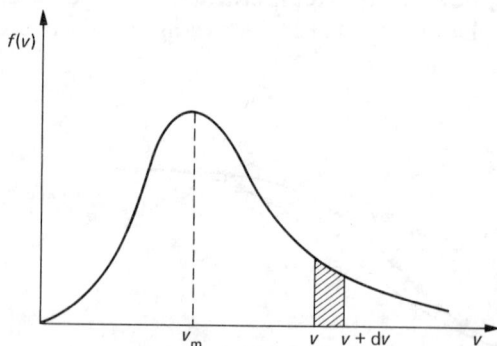

Plot for fixed β and fixed T

Fig. 9.15

if molecules were picked at random this speed would be found most often. Now, differentiating we have

$$df/dv = Ae^{-\beta v^2}(2v - 2\beta v^3)$$
$$= 2Ae^{-\beta v^2}v(1 - \beta v^2)$$

and this is zero for $v = 0$, $v = \infty$ and $v = 1/\sqrt{\beta}$. The first two roots show that the curve touches the v axis at the origin and is asymptotic to it at infinity, while the other root gives $v_m = \sqrt{2kT/m}$.

The average speed of a molecule is

$$\bar{v} = \frac{\int_0^\infty Av^3 e^{-\beta v^2} dv}{\int_0^\infty Av^2 e^{-\beta v^2} dv}$$

Letting $\beta v^2 = \frac{1}{2}t^2$, we have $v = t/\sqrt{2\beta}$ and so $2v\beta dv = t dt$. Thus

$$dv = \frac{t dt}{2v\beta} = \frac{t dt}{2\beta t/\sqrt{2\beta}} = \frac{dt}{\sqrt{2\beta}}$$

Substituting for v and dv in our expression for average velocity we have

$$\bar{v} = \frac{\int_0^\infty A[t^3/(2\beta)^{3/2}]e^{-t^2/2}\,dt/\sqrt{2\beta}}{\int_0^\infty A(t^2/2\beta)e^{-t^2/2}\,dt/\sqrt{2\beta}}$$

We now use the following formula for integrating by parts:

$$\int u\omega dx = u \int \omega dx - \int \left(\int \omega dx\right) \frac{du}{dx}\,dx$$

where u and ω are both functions of x.

To evaluate the numerator in our expression for $\bar{v}$ we write

$$I = \int_0^\infty t^3 e^{-t^2/2} dt = \int_0^\infty t^2 . t e^{-t^2/2} dt$$

and, letting t^2 represent u and $te^{-t^2/2}$ represent ω in the formula for integrating by parts, we have

$$I = t^2 \int_0^\infty te^{-t^2/2} dt - \int_0^\infty \left(\int te^{-t^2/2} dt\right) 2t dt$$

$$= t^2 \int_0^\infty e^{-t^2/2} d(\tfrac{1}{2}t^2) - \int_0^\infty \left(\int e^{-t^2/2} d(\tfrac{1}{2}t^2)\right) 2t dt$$

$$= \left[-t^2e^{-t^2/2}\right]_0^\infty - \int_0^\infty - e^{-t^2/2}.2t\,dt$$

$$= \left[-t^2e^{-t^2/2}\right]_0^\infty + 2\int_0^\infty te^{-t^2/2}\,dt$$

$$= 0 + 2\left[-e^{-t^2/2}\right]_0^\infty$$

$$= 2$$

To evaluate the denominator in our expression for $\bar{v}$ we write

$$I = \int_0^\infty t^2e^{-t^2/2}\,dt = \int_0^\infty t.te^{-t^2/2}\,dt$$

and this time we let $t \equiv u$ and $te^{-t^2/2} \equiv \omega$. Thus

$$I = \left[t(-e^{-t^2/2})\right]_0^\infty - \int_0^\infty -e^{-t^2/2}\,dt$$

$$= 0 + \tfrac{1}{2}\sqrt{2\pi}$$

since the second term is a standard integral.

Therefore, we have

$$\bar{v} = \frac{A(2\beta)^{-3/2}(2\beta)^{-1/2}.2}{A(2\beta)^{-1}(2\beta)^{-1/2}.\tfrac{1}{2}\sqrt{2\pi}} = \frac{1}{(2\beta)^{1/2}}\frac{4}{\sqrt{2\pi}}$$

$$= \left(\frac{16}{(2m/2kT)2\pi}\right) = \sqrt{\frac{8kT}{m\pi}}$$

The mean square velocity is given by

$$\overline{v^2} = \frac{\int_0^\infty Av^4e^{-\beta v^2}\,dv}{\int_0^\infty Av^2e^{-\beta v^2}\,dv}$$

But the average kinetic energy of a molecule is $\tfrac{1}{2}m\overline{v^2} = \tfrac{3}{2}kT$. Therefore $\overline{v^2} = 3kT/m$, and the root mean square (RMS) velocity $\sqrt{(\overline{v^2})} = \sqrt{(3kT/m)}$. Thus we have in order of increasing magnitude $v_m : \bar{v} : \sqrt{(\overline{v^2})} = \sqrt{2} : \sqrt{(8/\pi)} : \sqrt{3}$. With reference to Fig. 9.16,

$$v_m = \sqrt{\tfrac{2}{3}}\,v_{rms} = 0.817\,v_{rms}$$

$$\bar{v} = \sqrt{\tfrac{8}{3\pi}}\,v_{rms} = 0.921\,v_{rms}$$

and

$$v_{rms} = \sqrt{\frac{3RT}{M}}$$

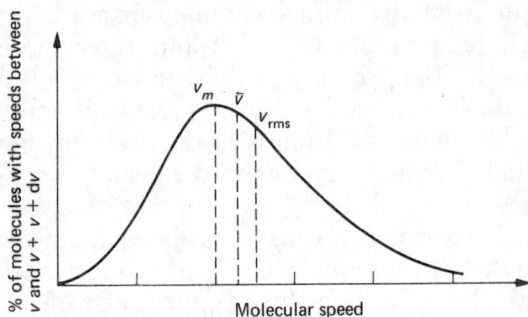

Fig. 9.16

where R is the universal gas constant and M the molecular weight of the gas. As the temperature T rises, $1/kT$ falls and the shape of the curve changes. The peak moves progressively to the right, for its position is given by $v_m = \sqrt{2kT/m}$ and it simultaneously decreases in height, the distribution becoming more spread out. The total area under the curve remains the same, since it represents the total number of molecules.

Experimental test of Maxwell's law

Fig. 9.17 shows the apparatus used by Zartman and Ko (1930) in order to test Maxwell's law experimentally. Bismuth mole-

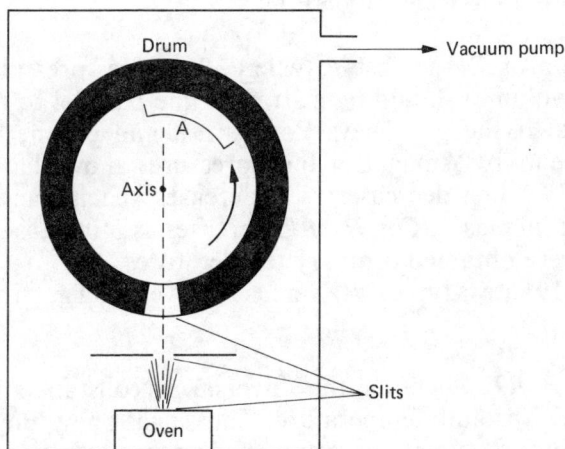

Fig. 9.17

cules stream from the oven containing bismuth vapour at a known high temperature (827 °C) and enter the evacuated region above it. The presence of a slit above the oven, in addition to the slit in the oven, collimates the beam of molecules, and when the slit in the drum is in the position shown in the diagram and the drum is stationary, the beam is deposited on a glass plate A.

During the experiment the drum is rotated at constant angular velocity, so that once during each rotation when the drum slit is aligned with the other two slits, a short burst of molecules is admitted. Since the molecules are travelling with different speeds, some cross the diameter more quickly than others, and owing to the rotation of the drum they strike the cold glass plate A at different places. Hence the apparatus translates a speed distribution into a distribution in space around the inside of the drum, as indicated by the darkening of the glass where the bismuth is deposited. The density of deposit is measured optically, and gives good agreement with Maxwell's law. Since the experiment is carried out under vacuum, there are few gas molecules in the apparatus to cause collisions which would alter the magnitude and direction of the bismuth molecular velocities, and thereby confuse the result.

Fig. 9.18 shows the Maxwell–Boltzmann distribution of molecular velocities in the gases oxygen and hydrogen.

9.12   Deviations from the gas laws

The graph of PV against P (where P denotes pressure and V denotes volume) should be a straight line parallel to the pressure axis, if the gas obeys Boyle's law and is a perfect gas. Experiments by Amagat at high pressures show that for air, CO_2, N_2, PV first decreases as P increases, reaches a minimum and then increases. For H_2, PV increases as P increases. These curves were obtained ordinary temperatures.

Fig. 9.19 shows typical PV versus P curves for a gas. In general

$$PV = A + BP + CP^2 + \ldots$$

where $A = RT$, where R is the universal gas constant, and where T denotes absolute temperature. Thus a gas obeys Boyle's law at very low pressures, where only the first term is of significance. From experience C and D etc. are always positive, but for all

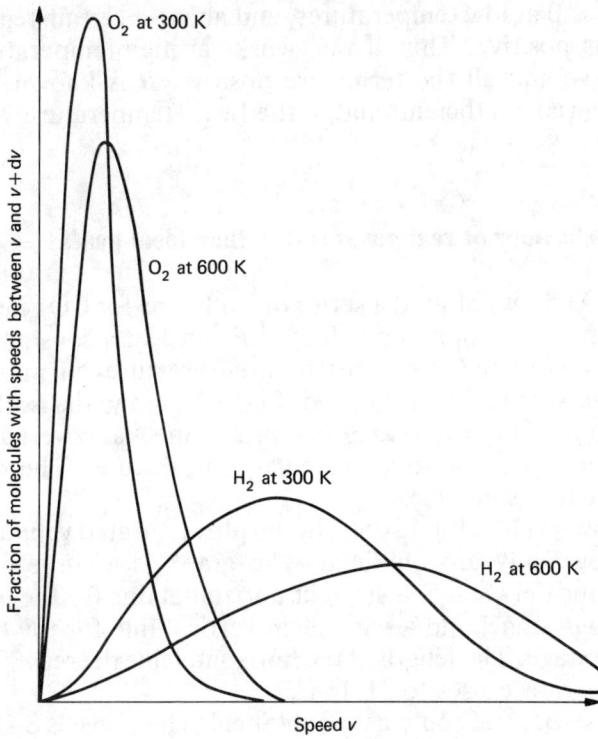

Fig. 9.18

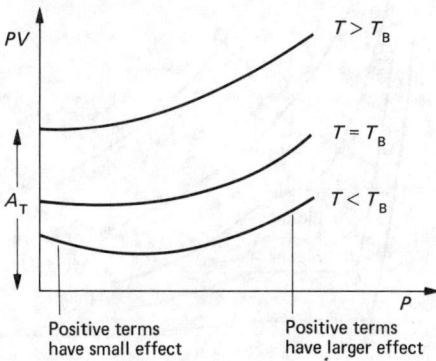

Fig. 9.19

gases $B < 0$ at low temperatures, and above a certain temperature B is positive. Thus if one works at high temperatures B is positive and all the terms are positive. B is known as the second virial coefficient, and at the Boyle temperature T_B, B is zero.

9.13   Behaviour of real gases (other than ideal gases)

In 1863 Andrews plotted a series of isotherms for CO_2, i.e. each curve gave corresponding values of P and V at a constant temperature. With the apparatus he used pressures of up to 100 atmospheres could be produced. Fig. 9.20 shows the isotherms obtained for CO_2 (i.e. a range of temperature was covered, each curve corresponding to a particular temperature). The results obtained are listed below:

(1) Below $31 \cdot 1$ °C liquid only, liquid plus saturated vapour, and vapour only are obtained. The graph is a curve for the vapour-only stage, a straight horizontal line for the liquid–vapour stage, and a very steep vertical line for the liquid-only stage. The length of the horizontal line decreases as the temperature rises to $31 \cdot 1$ °C.

(2) Above $31 \cdot 1$ °C only gas is obtained. The curve is S-shaped when a few degrees above $31 \cdot 1$ °C, but at higher temperatures it becomes a smooth curve.

The deduction made from this experiment is that above $31 \cdot 1$ °C CO_2 cannot be liquefied, however great the pressure.

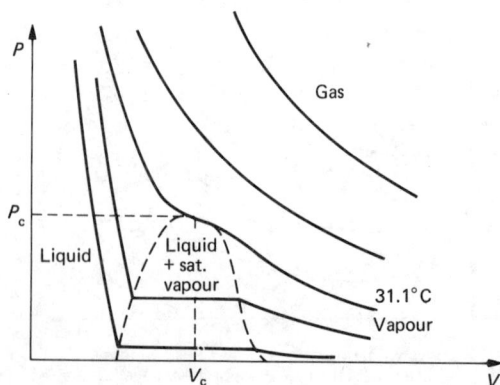

Fig. 9.20

This temperature is called the critical temperature of the gas, and may be defined as the temperature above which a gas cannot be liquefied by pressure alone. At the critical temperature the gas is just liquefied at the critical pressure P_c. The point $P_c V_c$ is the critical point, and the critical pressure is the pressure at the critical point. The critical volume V_c is the volume of one kilogram of a substance at the critical temperature and pressure.

Thus a gas may be defined as a substance in the gaseous state, above its critical temperature, while a vapour may be defined as a substance in the gaseous state below its critical temperature.

van der Waals' equation
van der Waals attempted to find an equation of state for real gases. He modified the ideal gas equation $PV = RT$ by amending two of the simplifying assumptions. First, the molecules cannot be infinitely small. Assume that the molecules are hard spheres of radius r: the centre of a molecule can never approach within a distance r of the wall of a containing vessel, or a distance $2r$ of the centres of the other molecules. Thus the volume available to a molecule is smaller than the real volume V (volume of container). Hence we must subtract from the real volume V, the co-volume b. Thus the ideal gas equation is first modified to $P(V - b) = RT$, where b is a constant depending on the mass of the gas and the nature of the gas.

Secondly, van der Waals assumed that attractive forces exist between the molecules. Such forces decrease rapidly with increasing distance, hence any molecule is attracted by all other molecules lying within a certain small distance of it. Thus molecules lying well away from the walls of a containing vessel experience no net force since they are attracted equally by molecules all around them. However, molecules near the walls experience a net attractive force towards the interior of the gas because they are not completely surrounded by other molecules (see Fig. 9.21). Thus the measured pressure P is too low. Hence we write $(P + P')(V - b) = RT$, where P' is a pressure correction applied to P.

As a first approximation, the average force exerted on a molecule in the surface layer is proportional to n, where n is the average number of molecules per unit volume. Actually the layer near the wall is slightly depopulated. The total force exerted on all the molecules in one unit area of surface layer as a first approximation is also proportional to n. Hence $P' \propto n^2$.

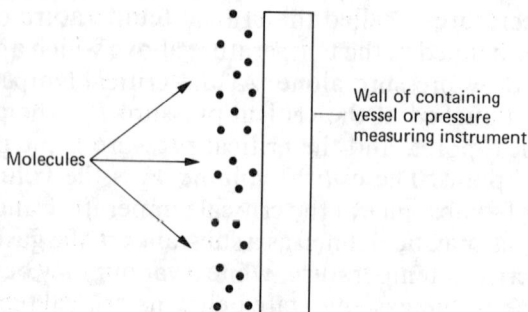

Fig. 9.21

But for one kilogram mole, $n = N_0/V$ where N_0 is Avogadro's number. Hence we let $P' = a/V^2$ where a is a constant, and have for van der Waals' equation

$$P + \frac{a}{V^2} (V - b) = RT$$

The constant b is proportional to the mass of gas taken, but a is proportional to the square of the mass. If V and R refer to one kilogram mole, i.e. R is the universal gas constant, then a and b refer to one kilogram mole.

The properties of van der Waals' equation
Referring to Fig. 9.22, if van der Waals' equation is multiplied out we get a cubic in V, and if one guesses suitable values for a and b the equation may be solved. Below a certain value of temperature T, T_c, three roots are obtained. Above T_c only

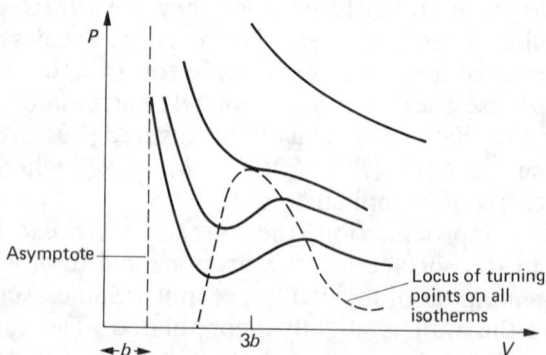

Fig. 9.22

one real solution is obtained and two imaginary solutions of no physical significance, i.e. for any *P* there is only one value of *V*. Since

$$P = \frac{RT}{V - b} - \frac{a}{V^2}$$

P becomes infinite if $V = b$.

To find the turning points (maxima and minima) on the isotherms we differentiate obtaining

$$\frac{\partial P}{\partial V}_T = \frac{2a}{V^3} - \frac{RT}{(V - b)^2}$$

For a turning point $(\partial P/\partial V)_T = 0$, thus

$$\frac{RT}{V - b} = \frac{2a(V - b)}{V^3}$$

Eliminating *T* from this equation and

$$P = \frac{RT}{V - b} - \frac{a}{V^2}$$

gives

$$P = \frac{a(V - 2b)}{V^3}$$

as the locus of the turning points on all the isotherms. This has a maximum when $V = 3b$ (differentiate with respect to *V* and equate to zero).

Points of inflexion are characterized by the gradient of the curve being a maximum or minimum, and so for a point of inflexion we have $(\partial^2 P/\partial V^2)_T = 0$. Now,

$$\frac{\partial^2 P}{\partial V^2}_T = \frac{2RT}{(V - b)^3} - \frac{6a}{V^4} = 0$$

when

$$RTV^4 = 3a(V - b)^3$$

Combining this with the equation

$$\frac{RT}{V - b} = \frac{2a(V - b)}{V^3}$$

there is only one point on the whole diagram at which there is an inflexion, and this is given by $V = \frac{3}{2}(V - b)$, so that $V = 3b$.

From Andrews' experiment this point is naturally identified with the critical point. Also, instead of the horizontal sections of the curves in Andrews' experiments, there is a region where $\partial P/\partial V > 0$ and this corresponds to the liquid plus vapour region.

Now assume that some fluid obeys van der Waals' equation and that the critical volume V_c, the critical temperature T_c, and the critical pressure P_c correspond to the point of inflexion. We will obtain a and b from P_c, V_c, T_c:

$$P + \frac{a}{V^2} = \frac{RT}{V - b}$$

Thus

$$\frac{dP}{dV} - \frac{2a}{V^3} = -\frac{RT}{(V - b)^2}$$

(T is constant along an isotherm). We now use the condition $dP/dV = 0$ for a point of inflexion. Thus

$$\frac{2a}{V_c^3} = \frac{RT_c}{(V_c - b)^2} \tag{1}$$

Differentiating again we obtain

$$\frac{d^2P}{dV^2} + \frac{6a}{V^4} = \frac{2RT}{(V - b)^3}$$

But the second derivative is also zero at a point of inflexion. Thus

$$\frac{3a}{V_c^4} = \frac{RT_c}{(V_c - b)^3} \tag{2}$$

Equations (1) and (2) represent the point of inflexion, and dividing Equ. (1) by Equ. (2) to eliminate T, we obtain $\frac{2}{3}V_c = V_c - b$ i.e. $V_c = 3b$. Putting $3b$ for V_c in Equ. (1) gives

$$\frac{2a}{27b^3} = \frac{RT_c}{4b^2}$$

and hence

$$RT_c = \frac{8a}{27b}$$

For the pressure

$$P + \frac{a}{V^2} = \frac{RT}{V - b} \quad \text{and} \quad P_c = \frac{8a}{27b}\frac{1}{2b} - \frac{a}{9b^2} = \frac{a}{9b^2}(\tfrac{4}{3} - 1)$$

Therefore $P_c = a/27b^2$, and as we have seen

$$RT_c = \frac{8a}{27b} \quad \text{and} \quad V_c = 3b$$

Assuming that these critical quantities can be measured any two give a and b, and assuming that the point of inflexion corresponds to the critical point, van der Waals' equation is the best that can be done as a treatment, without taking account of the forces between the molecules and of how they behave when close together. When van der Waals' curves are superimposed on the isothermals for CO_2 say, there is found to be an error of between 30 and 40%. The constants a and b vary considerably with temperature and volume in ways which cannot be forseen. One would expect b to vary with temperature, since the harder the molecules hit one another the more they interpenetrate.

Tests for van der Waals' equation

DIMENSIONLESS TEST
$RT/PV = 1$ for an ideal gas. On van der Waals' theory RT_c/P_cV_c should be $\frac{8}{3} = 2 \cdot 667$. Actually what one obtains for different gases varies considerably from $3 \cdot 2$ to $3 \cdot 8$, so that although the theory is considerably better than the ideal gas theory, it is still only an approximation.

SECOND VIRIAL COEFFICIENT TEST
As we have seen $PV = A + BP + CP^2 + \ldots$ where B is the second virial coefficient. Now van der Waals' equation is

$$P + \frac{a}{V^2} \ (V - b) = RT$$

and if we multiply this out neglecting ab/V^2 (small correcting term), and putting $a/V = aP/RT$ (approximately correct for moderate pressure and volume), we obtain

$$PV - Pb + \frac{aP}{RT} = RT$$

Thus

$$PV = RT + P \ b - \frac{a}{RT} \quad \text{and} \quad b - \frac{a}{RT}$$

is van der Waals' expression for the second virial coefficient. We see that this changes sign: being negative at low temperatures and positive at high temperatures. The Boyle temperature T_B is given by $T_B = a/Rb$. But $T_c = 8a/27Rb$, thus T_B/T_c should equal $27/8 = 3 \cdot 375$. Experimentally this varies from gas to gas between $2 \cdot 5$ and $3 \cdot 6$.

Historically a was obtained by a method other than the critical-constant method. Either the temperature coefficient of pressure at constant volume, k_V, or the temperature coefficient of volume at constant pressure, k_p, may be used. Since, however, the use of k_p is much more complicated, we shall consider the use of k_V. For a real gas

$$\frac{P - P_0}{P_0} = k_V (T - 273)$$

where P denotes pressure, the suffix 0 denotes the ice point and T denotes absolute temperature. This is an arbitrary definition of something which can be measured. Thus

$$P = P_0 [1 + k_V (T - 273)]$$

But

$$P + \frac{a}{V_c^2} = \frac{RT}{V_c - b}$$

where c denotes constant volume, and for the ice point

$$P_0 + \frac{a}{V_c^2} = \frac{273R}{V_c - b}$$

Subtracting these two equations we have

$$P - P_0 = \frac{R}{V_c - b} (T - 273)$$

But

$$R = \left(P_0 + \frac{a}{V_c^2} \right) \frac{V_c - b}{273}$$

Therefore

$$P - P_0 = \left(P_0 + \frac{a}{V_c^2} \right) (T - 273) \frac{1}{273}$$

k_V then follows as

$$k_V = \frac{1}{273} \left(1 + \frac{a}{P_0 V_c^2} \right)$$

and measurement of k_V gives a. For an ideal gas there are no forces between the molecules, $k_V = \frac{1}{273}$ and Charles' law is obtained. V_c is the volume one mole of gas is chosen to occupy. If V_c is large the gas is getting rare, and the term $a/P_0 V_c^2$ is small. Thus van der Waals' equation can be tested for gases of different densities.

The law of corresponding states
The constants a and b in van der Waals' equation may be eliminated by substituting for them expressions derived from the values of the critical constants given by $V_c = 3b$, $RT_c = 8a/27b$, and $P_c = a/27b^2$. We then obtain an equation which contains only the ratios of the pressure, volume and temperature, to their critical values. These ratios are called the reduced values of the variables. They may be written

$$\phi = \frac{V}{V_c} \quad \theta = \frac{T}{T_c} \quad \pi = \frac{P}{P_c}$$

and we can rewrite

$$\left(P + \frac{a}{V^2}\right)(V - b) = RT$$

as

$$\left(\pi \frac{a}{27b^2} + \frac{a}{\phi^2 . 9b^2}\right)(\phi.3b - b) = \theta RT_c$$

$$= \theta \frac{8a}{27b}$$

Thus

$$\left(\frac{\pi}{27} + \frac{1}{9\phi^2}\right)(3\phi - 1) = \frac{8\theta}{27}$$

or

$$\left(\pi + \frac{3}{\phi^2}\right)(3\phi - 1) = 8\theta$$

This equation contains no explicit reference to either the quantity, or the nature of the substance, so that it should apply in this form to all fluids. In fact van der Waals' equation is not accurately true for any known fluid, hence the last equation is not universally true.

Nevertheless it embodies a useful concept, the law of corresponding states. According to this when reduced pressure, volume and temperature are used, if the values of any pair of these variables are equal for two fluids, the two substances are in corresponding states and the third variable will have the same value for each. Any equation of state which contains only two constants, in addition to the gas constant R, leads to the law of corresponding states.

9.14   The Joule–Thomson or Joule–Kelvin effect

With reference to Fig. 9.23, the apparatus contains two frictionless pistons and a porous plug, usually made of porous earthenware. A gas is allowed to expand from a high pressure P_1, to a low pressure P_2, and to do work at the expense of its own internal energy, the gas seeping through the porous plug. The gas is thereby cooled; the effect is used for reaching low temperatures and for liquefying oxygen and nitrogen.

Imagine that the two pressures remain constant by allowing the left-hand piston to move forward and the right-hand piston to move to the right. Assume that one mole of gas seeps through the plug. The left-hand piston moves through the volume occupied by the mole of gas at the high pressure P_1, and the right-hand piston moves back to make up for the volume occupied by the mole of gas at the low pressure P_2.

Let V_1 be the volume per mole at pressure P_1, and let V_2 be the volume per mole at the pressure P_2. We now use van der Waals' equation, which gives $(P + a/V^2)(V - b) = RT$. Ignor-

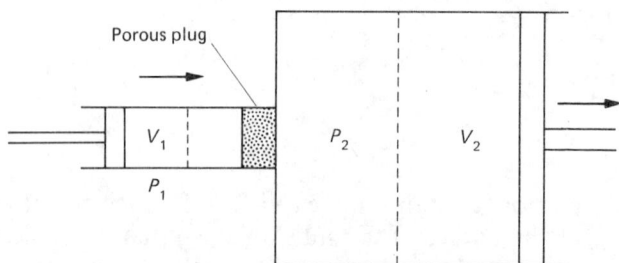

Porous plug

V_1

P_1

P_2

V_2

Fig. 9.23

ing ab, and putting $a/V = aP/RT$ approximately, we have

$$PV = RT + P\left(b - \frac{a}{RT}\right)$$

Now the net external work done by us is $P_1V_1 - P_2V_2$, the second term being the work done by the gas (work done = $\int PdV$). But the back drag at the wall $P' = a/V^2$ per unit area, and as the mole passes from the smaller volume to the larger volume, it does work against this. Thus the internal work done by the gas, according to van der Waals, is

$$\int_{V_1}^{V_2} \frac{a}{V^2} dV = \left[-\frac{a}{V}\right]_{V_1}^{V_2}$$

$$= -\left[\frac{a}{V_2} - \frac{a}{V_1}\right]$$

If the gas were ideal (no attactive forces) no such work would be done. The total work done by us is then

$$P_1V_1 - P_2V_2 + \frac{a}{V_2} - \frac{a}{V_1}$$

Since it is difficult to calculate the work done if the temperature changes, we arrange an exchange of heat in order to keep the temperature constant. Now using

$$PV = RT + P\left(b - \frac{a}{RT}\right)$$

the work done by us is equal to

$$(P_1 - P_2)\left(b - \frac{a}{RT}\right) + \frac{a}{RT}(P_2 - P_1)$$

$$= (P_1 - P_2)\left(b - \frac{2a}{RT}\right)$$

and the heat removed to keep the temperature constant is is $C_p\Delta T$, where C_p denotes specific heat at constant pressure and ΔT denotes temperature change. Thus

$$C_p\Delta T = (P_1 - P_2)\left(b - \frac{2a}{RT}\right)$$

Depending on the conditions the temperature may rise or

fall. If we want to use the effect to produce cooling, i.e. ΔT is negative, then the condition $2a/RT > b$ must apply. Hence to produce any cooling at all the temperature must be below a certain value T_i; If not a temperature rise occurs. T_i is called the inversion temperature, and is given by $T_i = 2a/Rb$. But the Boyle temperature $T_B = a/Rb$. Thus as a check on van der Waals' equation, we expect that $T_i = 2 T_B$. A rough agreement with this relationship is indeed found.

Table 9.3 shows that there is the usual numerical difference between theory and experiment for real gases. For H_2 and He the inversion temperatures are low, and these gases must be pre-cooled in order to use the Joule–Kelvin effect. All other gases except H_2 and He have much higher inversion temperatures (above room temperature). The magnitude of the effect is very dependent on the pressure difference. In industry, pressure differences of 150 atmospheres are used to produce a larger effect.

Table 9.3   Experimental results for the inversion and Boyle temperatures.

	T_B/K	T_i/K	T_i/T_B
H_2	106	193	1·82
He	19	30	1·58

10 The First Law of Thermodynamics

Before commencing our discussion of thermodynamics, and in particular of the first law, we will give a brief review of the mathematics required.

Thermodynamics is essentially restricted to conditions of equilibrium. We have a restricted number of independent variables, e.g. P (pressure), V (volume), T (temperature), energy, entropy, free energy etc., and simple systems have their states fixed by two variables.

Consider the implicit function $f(x, y) = 0$. We may write

$$\left(\frac{\partial f}{\partial x}\right)_y dx + \left(\frac{\partial f}{\partial y}\right)_x dy = 0$$

Therefore

$$\frac{dy}{dx} = -\frac{\partial f/\partial x}{\partial f/\partial y}$$

Now consider the implicit function $f(x, y, z) = 0$. It may be shown (see a standard mathematics text, and consult the section on partial differentiation) that

$$\left(\frac{\partial x}{\partial y}\right)_z \left(\frac{\partial y}{\partial z}\right)_x \left(\frac{\partial z}{\partial x}\right)_y = -1$$

In particular for P, V, T

$$\left(\frac{\partial P}{\partial V}\right)_T \left(\frac{\partial V}{\partial T}\right)_P \left(\frac{\partial T}{\partial P}\right)_V = -1$$

An exact differential
$Mdx + Ndy$ is an exact differential if there exists a function $f(x, y)$, such that the differential

$$df = \frac{\partial f}{\partial x} dx + \frac{\partial f}{\partial y} dy = Mdx + Ndy$$

For an exact differential the line integral

$$\int (M\mathrm{d}x + N\mathrm{d}y)$$

depends only on the end points, and not on the path of integration. A necessary and sufficient condition that $M\mathrm{d}x + N\mathrm{d}y$ should be an exact differential, is that

$$\frac{\partial M}{\partial y} = \frac{\partial N}{\partial x}$$

provided that these exist, are continuous and single valued. If $M\mathrm{d}x + N\mathrm{d}y$ is not exact it can be made exact by multiplying by an integrating factor.

We shall now deal with thermodynamics proper. For a system in an equilibrium thermodynamic state, there are no pressure differences, no expansions or contractions and no heat flow. The state is recognized by steady unchanging values of temperature T, pressure P and volume V etc associated with the system. It is an experimental fact that simple systems have their states fixed by only two state variables. When two of these have certain values, any other always has the same value. With reference to Fig. 10.1, the point represents a state and fixes any third variable, e.g. T. If a, b, c are three variables and we can find an analytic function $c = f(a, b)$, this is an equation of state. In particular $T = f(P, V)$, $P = f(T, V)$ and $V = f(T, P)$ are called characteristic equations. The best known is $PV = RT$ for a perfect gas. Other equations are only approximate.

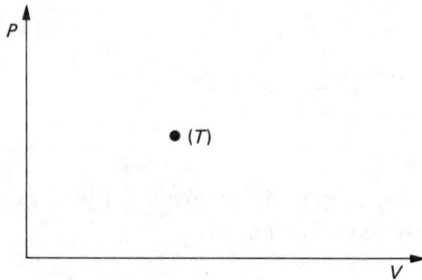

P

$\bullet$ (T)

V

Fig. 10.1

10.1 Internal energy and the first law of thermodynamics

Experimentally the heat given to a system is generally not equal to the work done. Thus the excess energy must have gone into the system, or the deficit in energy must have been extracted from the system. This gives rise to the concept of the internal energy of the system. The first law of thermodynamics may be stated as follows: a quantity called internal energy, U, exists, whose change of value is the difference between the heat energy which has entered the system and the mechanical work which has been done by the system, and which reverts to its original value when the system reverts to its original state. This means that if we do a series of experiments and come back to the original state, there is no change in U. Thus U represents the state (along with the other variables).

Algebraically, dealing with infinitesimal changes, $dQ = dU + dW$ per kilogram, where Q denotes heat and W denotes work. But U is a function of the state. Therefore $U = f(T, V)$, where the T part represents the vibration or motion of the molecules, and the V part represents the potential energy of the molecules. Therefore

$$dU = \left(\frac{\partial U}{\partial T}\right)_V dT + \left(\frac{\partial U}{\partial V}\right)_T dV$$

Therefore

$$\frac{dQ}{dT} = \left(\frac{\partial U}{\partial T}\right)_V + \left(\frac{\partial U}{\partial V}\right)_T \frac{dV}{dT} + \frac{dW}{dT}$$

Now dQ/dT is clearly specific heat, and for the specific heat at constant volume C_V there is no external work and no expansion. Thus the heat produces an increase in internal energy. Therefore in particular

$$C_V = \left(\frac{\partial U}{\partial T}\right)_V$$

In general

$$C = C_V + \left(\frac{\partial U}{\partial V}\right)_T \frac{dV}{dT} + \frac{dW}{dT}$$

The case of an ideal gas
This is simple for two reasons:
(1) We have the experimental result that U depends on temperature only. If a gas is indefinitely rarefied the molecules are so far apart that there are no attractive forces between them. Thus U depends only on the kinetic energy of the molecules. Therefore $(\partial U/\partial V)_T = 0$.

Fig. 10.2 shows diagrammatically the apparatus used in an experiment by Joule. A gas was allowed to expand into a vacuum. Thus no external work was done and no heat flowed in or out since the system was insulated. The first law gives $0 = dU + 0$. Therefore

$$\left(\frac{\partial U}{\partial T}\right)_V dT + \left(\frac{\partial U}{\partial V}\right)_T dV = 0$$

But dV is finite, and $dT = 0$ since Joule could detect no change in temperature. Therefore $(\partial U/\partial V)_T = 0$. In fact there is a small change in temperature for a real gas although Joule could not detect it, and $(\partial U/\partial V)_T = 0$ only for an ideal gas.

(2) For an ideal gas there is an equation of state, i.e. $PV = RT$ for one kilogram mole. Now $dQ = dU + PdV$ per kilogram mole (work $dW = PdV$). Thus

$$dQ = \left(\frac{\partial U}{\partial T}\right)_V dT + \left(\frac{\partial U}{\partial V}\right)_T dV + PdV$$

But $(\partial U/\partial V)_T = 0$ for an ideal gas. Therefore

$$dQ = C_V dT + PdV$$

since $(\partial U/\partial T)_V = C_V$. Thus

$$\frac{dQ}{dT} = C_V + P\frac{dV}{dT}$$

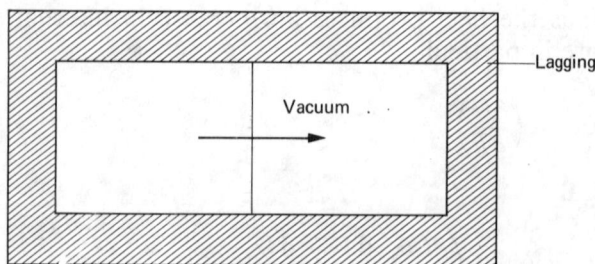

Fig. 10.2

At constant pressure $dQ/dT = C_p$, where C_p is the molar specific heat at constant pressure. Therefore, we have

$$C_p = C_V + P \left(\frac{\partial V}{\partial T} \right)_P$$

per kilogram mole. But $PV = RT$, and so

$$\left(\frac{\partial V}{\partial T} \right)_P = \frac{R}{P}$$

Therefore $C_p - C_v = R$ per kilogram mole.

10.2 Reversible processes

A reversible process is one that is in equilibrium at each instant, the system effectively going through a succession of equilibrium states. The process can be made to take place in the opposite sense by an infinitesimal change (reversal) in the conditions. No actual process is fully reversible, but many processes, when carried out slowly, are practically reversible.

Friction and eddies produce irreversible thermal and mechanical effects. Mechanical energy is converted into heat and in the reverse process this heat is not reconverted, but instead more mechanical energy is converted into heat. When a reversible process is carried out in the reverse direction, then the whole system goes through exactly the same series of changes in the reverse direction, e.g. P corresponding to a given V is always the same. For complete reversibility heat flows must also be exactly reversed if the process proceeds in reverse, and temperature differences must therefore be infinitesimal only. At the conclusion of a reversible process, both the system and its local surroundings may be returned to their initial states without producing any change in the universe.

Consider now a reversible transition from one state to another. For a gas, this can be represented by an equation connecting the three variables P, V, T.

(1) Consider a reversible isothermal process. This may be represented for an ideal gas by the Boyle's law relationship at a particular temperature which is kept constant, i.e. $PV = RT$ (T constant) represents a reversible isothermal process.

(2) Consider a reversible adiabatic process for an ideal gas.

The equation for an adiabatic change is obtained by stating the first law under conditions of no heat exchange, i.e. for one kilogram mole

$$0 = dU + PdV = C_V dT + PdV$$

Now, using $PV = RT$ we can write

$$0 = C_V dT + \frac{RT}{V} dV$$

But $C_p - C_V = R$ or $C_V(\gamma - 1) = R$, where $\gamma = C_p/C_V$. Thus

$$0 = dT + (\gamma - 1)T \frac{dV}{V}$$

and

$$\frac{dT}{T} + (\gamma - 1)\frac{dV}{V} = 0$$

Integrating, we find

$$\ln TV^{\gamma-1} = \text{constant}$$

or

$$TV^{\gamma-1} = \text{constant}$$

Substituting for T in terms of P and $V (PV = RT)$ we find

$$PVV^{\gamma-1} = \text{constant}$$

since R can be absorbed into the constant. That is

$$PV^\gamma = \text{constant}$$

Again, substituting for V in terms of T and P, we have

$$T\frac{T^{\gamma-1}}{P^{\gamma-1}} = \text{constant}$$

i.e.

$$\frac{T^\gamma}{P^{\gamma-1}} = \text{constant}$$

Thus a reversible adiabatic process for an ideal gas may be represented by $PV^\gamma = \text{constant}$, $TV^{\gamma-1} = \text{constant}$ or $T^\gamma/P^{\gamma-1} = \text{constant}$.

Work done by an ideal gas in a reversible adiabatic expansion
The work done, W, by an ideal gas in a reversible adiabatic expansion is given by

$$W = \int_{V_1}^{V_2} P dV$$

under the conditions $PV^\gamma = $ constant. But the first law gives

$$0 = dU + PdV$$
$$= C_V dT + PdV$$

Therefore

$$W = \int PdV = -C_V(T_2 - T_1)$$

The work is done at the expense of the internal energy, therefore the temperature drops and $T_2 < T_1$. Thus $W = C_V(T_1 - T_2)$. But $C_V(\gamma - 1) = R$. Therefore

$$W = \frac{R(T_1 - T_2)}{\gamma - 1} = \frac{P_1 V_1 - P_2 V_2}{\gamma - 1}$$

Work done by an ideal gas in a reversible isothermal expansion
The work done during an isothermal expansion is given by

$$\int PdV = \int_{V_1}^{V_2} \frac{RT}{V} dV = RT \ln (V_2/V_1)$$

10.3 Enthalpy

The enthalpy of a system is defined as $H = U + PV$. Consider the change in enthalpy which takes place when a system undergoes an infinitesimal process from an initial equilibrium state to a final equilibrium state:

$$dH = dU + PdV + VdP$$

But $dQ = dU + PdV$, therefore

$$dH = dQ + VdP$$

and

$$\frac{dH}{dT} = \frac{dQ}{dT} + V\frac{dP}{dT}$$

At constant pressure

$$\left(\frac{\partial H}{\partial T}\right)_P = \left(\frac{dQ}{dT}\right) = C_p$$

where C_p denotes specific heat at constant pressure.

Since $dH = dQ + VdP$ the change in enthalpy during a constant-pressure process is equal to the heat that is transferred:

$$H_f - H_i = Q$$

or

$$H_f - H_i = \int_i^f C_p \, dT$$

where i denotes initial state and f denotes final state. Also, since constant-pressure processes are much more important in engineering and chemistry than constant-volume processes, enthalpy is of greatest use in these branches of science. It is to be noted that since $H = U + PV$, and both U and PV have the dimensions of energy, then H behaves as energy and is sometimes called the total heat.

Referring to Fig. 10.3, the graph of Q against T is normally a smooth curve, the discontinuity indicating a phase change. The change in Q accompanying the phase change is defined as latent heat. Apply the first law of thermodynamics and consider only the end points rather than the process of the change, e.g. consider first a container full of liquid and then a container full of vapour. If there is no change in volume, no work will be done. Thus the

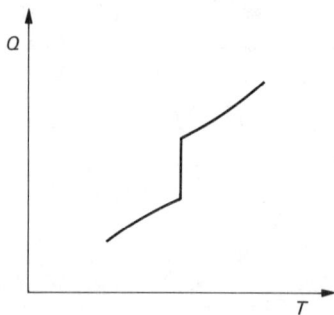

Fig. 10.3

latent heat at constant volume, $L_V = \Delta U = U_f - U_i$. This is not very useful as it is difficult to carry out a phase change at constant volume.

More useful is the change at constant pressure. In this case the first law gives $\Delta Q = \Delta U + P\Delta V = \Delta(U + PV)$, provided that the pressure is constant. Thus the latent heat at constant pressure, $L_p = \Delta H = H_f - H_i$. In the same way writing the first law for a chemical reaction, the heat of reaction is $H_i - H_f$, all at constant temperature. As an example of a problem on enthalpy consider the following.

Given that at 0 °C the latent heat of fusion of ice is $6039 \cdot 6 \times 10^3$ J mol^{-1} and the specific heat per mole is $75 \cdot 6 \times 10^3$ J K^{-1} for water and $37 \cdot 8 \times 10^3$ J K^{-1} for ice, what is the latent heat of fusion at -2 °C?

First, we have

$$6039 \cdot 6 \times 10^3 = H_{l0} - H_{s0}$$

where H_{l0} denotes the enthalpy of the liquid (water) at 0 °C and H_{s0} denotes the enthalpy of the solid (ice) at 0 °C. Using similar notation, we also have

$$H_{l-2} - H_{l0} = -2 \times 75 \cdot 6 \times 10^3 = -151 \cdot 2 \times 10^3$$
$$H_{s-2} - H_{s0} = -2 \times 37 \cdot 8 \times 10^3 = -75 \cdot 6 \times 10^3$$

Therefore

$$H_{l-2} - H_{s-2} = -151 \cdot 2 \times 10^3 + H_{l0} + 75 \cdot 6 \times 10^3 - H_{s0}$$
$$= -75 \cdot 6 \times 10^3 + 6039 \cdot 6 \times 10^3$$
$$= 5964 \cdot 0 \times 10^3 \text{ J mol}^{-1}$$

Thus the latent heat of fusion at -2 °C is $5964 \cdot 0 \times 10^3$ J mol^{-1}.

10.4 The Joule–Kelvin expansion (throttling process)

Referring to Fig. 10.4, the two pistons are moved simultaneously in such a way that a constant pressure P_i is maintained on the left-hand side of the plug and a constant lower pressure P_f is maintained on the right-hand side. The throttling process is irreversible, since the gas passes through non-equilibrium states on its way from the initial equilibrium state to the final equilibrium state. Considering these equilibrium states and applying

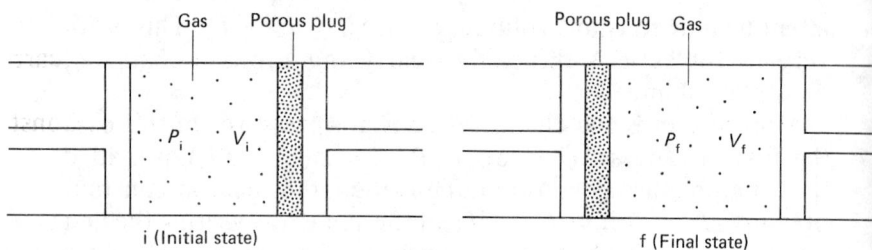

Fig. 10.4

the first law of thermodynamics

$$Q = U_f - U_i + W$$

But $Q = 0$ (because of thermal insulation and non-conducting pistons) and the work done is

$$W = \int_0^{V_f} P dV + \int_{V_i}^0 P dV$$

Since both pressures remain constant

$$W = P_f V_f - P_i V_i$$

and so

$$0 = U_f - U_i + P_f V_f - P_i V_i$$

i.e.

$$U_i + P_i V_i = U_f + P_f V_f$$

or

$$H_i = H_f$$

Thus $\Delta H = 0$ for a Joule–Kelvin expansion.

However, H is a function of the state $H(P, T)$. Therefore

$$\left(\frac{\partial P}{\partial T}\right)_H \left(\frac{\partial T}{\partial H}\right)_P \left(\frac{\partial H}{\partial P}\right)_T = -1$$

and since $(\partial H / \partial T)_P = C_p$,

$$\left(\frac{\partial T}{\partial P}\right)_H = -\frac{1}{C_p} \left(\frac{\partial H}{\partial P}\right)_T$$

and

$$\Delta T = -\frac{1}{C_p} \int \left(\frac{\partial H}{\partial P}\right)_T dP$$

where ΔT is the temperature change involved when a gas undergoes a Joule–Kelvin expansion.

Now $H = U + PV$ and for a perfect gas $(\partial U/\partial P)_T = 0$. Also for a perfect gas PV is independent of P since it equals RT. Thus $(\partial H/\partial P)_T$ is zero for a perfect gas, and there should be no temperature change when such a gas undergoes a Joule–Kelvin expansion. The temperature change ΔT for any real gas may be taken as a measure of the deviation of the gas from the ideal. Some non-ideal gases however, may also show a zero Joule–Kelvin effect but the reasons for this will not be discussed here. For air, O_2 and $N_2 \Delta T$ is negative, while for H_2 and $He \Delta T$ is positive (the starting temperature in all cases being room temperature).

Again, writing $H = U + PV$

$$dH = dU + PdV + VdP = dQ + VdP$$

But as we shall see later on when we discuss entropy S, dQ may be written TdS. Thus $dH = TdS + VdP$, and according to the second TdS equation

$$TdS = C_p dT - T\left(\frac{\partial V}{\partial T}\right)_P dP$$

(for the derivation of this equation, consult a more advanced book on thermodynamics). Substituting for TdS we get

$$dH = C_p dT - \left[T\left(\frac{\partial V}{\partial T}\right)_P - V\right]dP$$

and

$$dT = \frac{1}{C_p}\left[T\left(\frac{\partial V}{\partial T}\right)_P - V\right]dP + \frac{1}{C_p}dH$$

Also regarding T as a function of P and H,

$$dT = \left(\frac{\partial T}{\partial P}\right)_H dP + \left(\frac{\partial T}{\partial H}\right)_P dH$$

Thus, since the Joule–Kelvin coefficient $\mu = (\partial T/\partial P)_H$, we have

$$\mu = \frac{1}{C_P}\left[T\left(\frac{\partial V}{\partial T}\right)_P - V\right]$$

$$= -\frac{1}{C_P}(V - TV\beta)$$

where β, the coefficient of volume expansion of the gas, is $(1/V)(\partial V/\partial T)_P$, and so

$$\mu = -\frac{V}{C_P}(1 - \beta T)$$

At an inversion point μ is zero and we have

$$1 - \beta T_i = 0 \quad \text{or} \quad T_i = 1/\beta$$

where T_i is the inversion temperature.

Finally, returning to our previous expression for μ, we have

$$\mu = \left(\frac{\partial T}{\partial P}\right)_H = -\frac{1}{C_P}\left(\frac{\partial H}{\partial P}\right)_T$$

$$= \frac{1}{C_P}\left[-\left(\frac{\partial U}{\partial P}\right)_T - \left(\frac{\partial(PV)}{\partial P}\right)_T\right]$$

We see that the first term represents the deviation from Joule's law and the second term the deviation from Boyle's law.

Referring to Fig. 10.5, an isenthalpic curve is not the graph of a throttling process (a throttling process is irreversible), but is

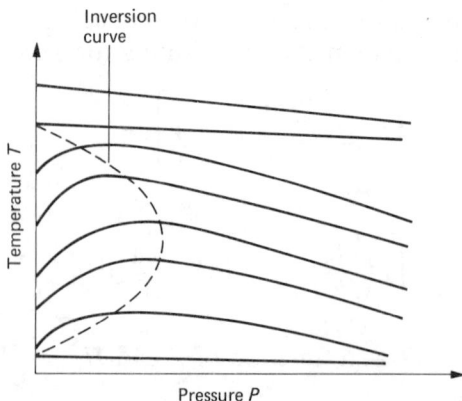

Fig. 10.5

the locus of all points representing equilibrium states of the same enthalpy. To obtain an isenthalpic curve $P_i T_i$ is kept fixed, and P_f set to any value less than P_i (see Fig. 10.4). T_f is then measured, and the procedure repeated for another value of P_f. The value of $P_i T_i$ is then changed to get another isenthalpic curve. The locus of all points at which the Joule–Kelvin coefficient μ is zero, i.e. the locus of the maxima of the isenthalpic curves, is known as the inversion curve. The region inside the inversion curve where μ is positive is called the region of cooling, whereas outside, where μ is negative, is the region of heating.

The use of the Joule–Kelvin expansion in establishing the Kelvin temperature scale

The Joule–Kelvin expansion gives a powerful method of establishing the Kelvin scale in the temperature range in which the Joule–Kelvin experiment can be performed. Consider θ a temperature measured on any scale, and let C_P^* be the specific heat at constant pressure measured on the θ scale:

$$C_P^* = \left(\frac{dQ}{d\theta}\right)_P = \left(\frac{dQ}{dT}\right)_P \left(\frac{\partial T}{\partial \theta}\right)_P = C_P \left(\frac{\partial T}{\partial \theta}\right)_P$$

where C_P is measured on the Kelvin scale. But $(\partial T/\partial \theta)_P = dT/d\theta$ since the derivative is independent of P. Thus

$$\frac{C_P^*}{C_P} = \frac{dT}{d\theta} = \frac{dP}{d\theta}\frac{dT}{dP}$$

and

$$C_P \frac{dT}{dP} = C_P^* \frac{d\theta}{dP}$$

Now, using $dT = \mu dP$, where μ is the Joule–Kelvin coefficient, we have

$$C_P \frac{dT}{dP} = T\left(\frac{\partial V}{\partial T}\right)_P - V$$

and

$$C_P^* \frac{d\theta}{dP} = T\left(\frac{\partial V}{\partial \theta}\right)_P \frac{d\theta}{dT} - V$$

Thus

$$\frac{\mathrm{d}T}{T} = \frac{(\partial V/\partial\theta)_P \mathrm{d}\theta}{C_P^*(\mathrm{d}\theta/\mathrm{d}P) + V}$$

and integrating we obtain

$$\log_e \frac{T_2}{T_1} = \int_{T_1}^{T_2} \frac{(\partial V/\partial\theta)_P \mathrm{d}\theta}{C_P^*(\mathrm{d}\theta/\mathrm{d}P) + V}$$

Looking at the expression on the right, $(\partial V/\partial\theta)_P$ can be obtained from the volume expansion coefficient in terms of empirical temperature. C_P^* is measured using the θ temperature scale, and $\mathrm{d}\theta/\mathrm{d}P$ is the inverse of change of pressure with empirical temperature. Thus all the quantities inside the integral sign can be measured between two temperatures, e.g. the ice point and the melting point of tin. A graph of

$$\frac{(\partial V/\partial\theta)_P}{C_P^*(\mathrm{d}\theta/\mathrm{d}P) + V}$$

against θ is then plotted, and the integration carried out graphically to give $\log_e T_2/T_1$. Theoretically this is the most important way of establishing the Kelvin scale in the range over which gases can be used.

10.5   Adiabatic and isothermal processes

With reference to Fig. 10.6 it may not always be possible to record the state of a body on a PV diagram, e.g. if a fluid is in a state of turbulence more than one P is required to describe it.

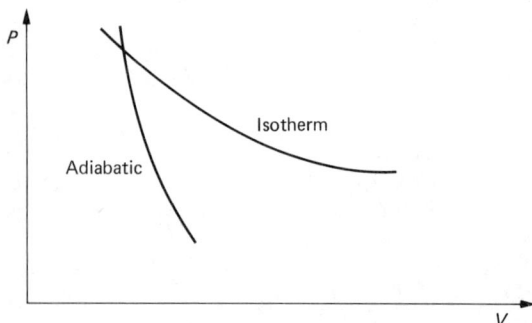

Fig. 10.6

However, we consider substances in equilibrum, i.e. we shall consider reversible processes. During a reversible adiabatic process the entropy (we shall discuss the concept of entropy later on) of a system remains constant, or in other words the system undergoes an isentropic process.

Consider an isothermal process and apply the first law:

$$dQ = dU + P\,dV$$

$$= \left(\frac{\partial U}{\partial T}\right)_V dT + \left(\frac{\partial U}{\partial V}\right)_T dV + P\,dV$$

The first term on the right-hand side is zero for an isotherm and hence

$$dQ = \left[\left(\frac{\partial U}{\partial V}\right)_T + P\right]dV$$

This is an expression for the heat taken in during an isothermal expansion, or the heat given out during an isothermal compression. For an ideal gas the heat transferred is equal to the work done since $(\partial U/\partial V)_T$ is zero in this case.

Now consider an adiabatic process and apply the first law:

$$0 = \left(\frac{\partial U}{\partial P}\right)_V dP + \left(\frac{\partial U}{\partial V}\right)_P dV + P\,dV$$

i.e. the work done equals the change in internal energy. Thus

$$0 = \left(\frac{\partial U}{\partial P}\right)_V dP + \left[\left(\frac{\partial U}{\partial V}\right)_P + P\right]dV$$

and, if the suffix S denotes constant entropy, we have

$$\left(\frac{\partial P}{\partial V}\right)_S = -\frac{\left[(\partial U/\partial V)_P + P\right]}{(\partial U/\partial P)_V}$$

$$= -\frac{\left[(\partial U/\partial T)_P(\partial T/\partial V)_P + P\right]}{(\partial U/\partial T)_V(\partial T/\partial P)_V}$$

$$= -\frac{(\partial T/\partial V)_P\left[(\partial U/\partial T)_P + P(\partial V/\partial T)_P\right]}{(\partial U/\partial T)_V(\partial T/\partial P)_V}$$

However,

$$\left(\frac{\partial U}{\partial T}\right)_V = C_V$$

$$\left(\frac{\partial U}{\partial T}\right)_P + P\left(\frac{\partial V}{\partial T}\right)_P = C_P$$

Thus

$$\left(\frac{\partial P}{\partial V}\right)_S = -\frac{C_P}{C_V}\frac{(\partial T/\partial V)_P}{(\partial T/\partial P)_V}$$

Now, since

$$\left(\frac{\partial T}{\partial V}\right)_P\left(\frac{\partial V}{\partial P}\right)_T\left(\frac{\partial P}{\partial T}\right)_V = -1$$

we have

$$\left(\frac{\partial P}{\partial V}\right)_S = \frac{C_P}{C_V}\left(\frac{\partial P}{\partial V}\right)_T$$

and, since $C_P > C_V$, this shows that a constant entropy curve is always steeper than an isothermal curve.

10.6   Summary of properties of U and H

Internal energy U

$$dU = dQ - PdV$$

$$\left(\frac{\partial U}{\partial T}\right)_V = C_V$$

Constant-volume process

$$\Delta U = Q$$
$$\Delta U = \int C_V dT$$

Adiabatic process

$$\Delta U = -\int PdV$$

Free expansion
$$\Delta U = 0$$

For an ideal gas

$$U = \int C_V dT + \text{constant}$$

Enthalpy H

$$dH = dQ + VdP$$

$$\left(\frac{\partial H}{\partial T}\right)_P = C_P$$

Constant-pressure process

$$\Delta H = Q$$
$$\Delta H = \int C_P dT$$

Adiabatic process

$$\Delta H = \int VdP$$

Joule–Kelvin expansion
$$\Delta H = 0$$

For an ideal gas

$$H = \int C_P dT + \text{constant}$$

11 The Second Law of Thermodynamics

The first law of thermodynamics allows the transformation of heat into work. Consider now a process by which the earth is cooled by $0 \cdot 1$ °C. Using the value for the mass of the earth, together with a rough value for its specific heat, a simple calculation shows that approximately 10^{20} kW h of energy would be released. When one considers that the annual UK consumption of electrical energy is approximately 2×10^{11} kW h, it is clear that the energy released by cooling the earth by $0 \cdot 1$ °C is very considerable.

Suppose now that in order to tap this energy we have a heat engine as shown in Fig. 11.1. This engine takes in heat Q, and performs an equivalent amount of work W. However, such an engine according to the second law of thermodynamics is impossible, otherwise for example a ship could be propelled across the Atlantic by extracting heat from the sea and converting this into useful work.

In practice a real heat engine always works between two reservoirs, one hot and the other cold. Heat Q_1 is absorbed from the hot reservoir, heat Q_2 is rejected to the cold reservoir, and work $W = Q_1 - Q_2$ is performed. For any heat engine a certain amount of heat is always rejected.

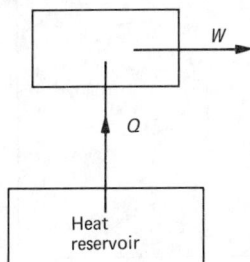

W

Q

Heat
reservoir

Fig. 11.1

11.1   The second law

Kelvin's statement: It is impossible for any periodically acting machine continuously to abstract heat from the coldest body of its surroundings, and convert this into useful mechanical work.

Clausius' statement: No self-acting machine, working in a cycle, can transfer heat continuously from a colder to a hotter body and produce no other external effect.

These two statements are in fact equivalent.

As an example of a problem on thermodynamics, we will show that $(\partial P/\partial V)_T$ must always be negative. Let a quantity of gas (or indeed any fluid) be confined in a closed vessel (see Fig. 11.2), and consider a small part A of this gas enclosed by an imaginary surface. If, owing to thermal fluctuations, this part of the gas should expand slightly, then if it is a normal gas with $(\partial P/\partial V)_T$ negative, its pressure will decrease, and the pressure of the surrounding gas will cause it to return to its initial volume. If, however, $(\partial P/\partial V)_T$ is positive, its pressure will increase when it expands, while the pressure of the rest of the gas will decrease. Thus the volume of A will increase more, while the remainder of the gas will contract. This process will go on until eventually there will be a phase of low density and a phase of much higher density; vapour and liquid.

Thus a gas for which $(\partial P/\partial V)_T$ is positive is bound to be unstable, and could not exist in a single phase form for more than a very short time before breaking up into two phases.

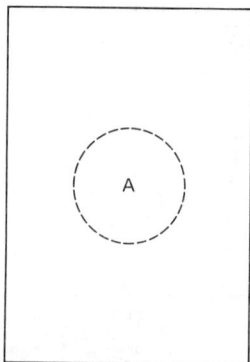

Fig. 11.2

11.2 The Carnot cycle

In Fig. 11.3 the Carnot cycle is defined as two reversible isothermal processes, AB (temperature T_1), CD (temperature T_2),
and two reversible adiabatic processes BC and DA. The reason
for this is that one wants all heat to be supplied reversibly at
one temperature and removed reversibly at a lower temperature, and one can only get reversibly from one temperature to
the other by using a reversible adiabatic expansion or contraction. During isothermal process AB heat Q_1 is absorbed from a
hot reservoir, and during isothermal process CD heat Q_2 is
rejected to a cold reservoir. At the end of process DA the working substance has returned to its original state. Hence $\Delta U = 0$
and the net heat in is $W = Q_1 - Q_2$. (Note that $W = \oint P\mathrm{d}V$ is
the area enclosed by the cycle on the $P-V$ diagram).

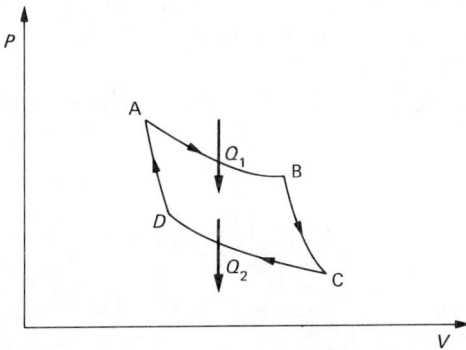

Fig. 11.3

Thus the Carnot cycle consists of a number of reversible processes, and for a process to be reversible: (*a*) it must be carried
out quasi-statically (the system effectively passes through a
series of equilibrium states); and (*b*) it must not be accompanied
by any dissipative effects (e.g. due to viscosity, friction, inelasti-
city, electric resistance, magnetic hysteresis).

The efficiency of a heat engine is given by

$$\frac{\text{net work done during one cycle}}{\text{heat absorbed at the higher temperature}}$$

$$= \frac{W}{Q_1} = \frac{Q_1 - Q_2}{Q_1} = 1 - \frac{Q_2}{Q_1}$$

With reference to Fig. 11.4, let the working substance be a gas for simplicity. Then the series of processes which form the Carnot cycle are as follows (see Figures 11.3 and 11.4):

(1) The cylinder is placed on the hot source and heat Q_1 is absorbed at temperature T_1. The gas expands isothermally from A to B performing work as it does so.

(2) The cylinder is placed on the insulating stand and the expansion allowed to continue adiabatically from B to C. More work is done and the temperature falls to T_2.

(3) The cylinder is placed on the cold sink, gives up heat Q_2, and contracts isothermally from C to D at temperature T_2. Work is done on the gas in this process.

(4) The cylinder is placed on the insulating stand and the compression continued adiabatically from D to A, the temperature rising to T_1.

Note that although we have considered the working substance

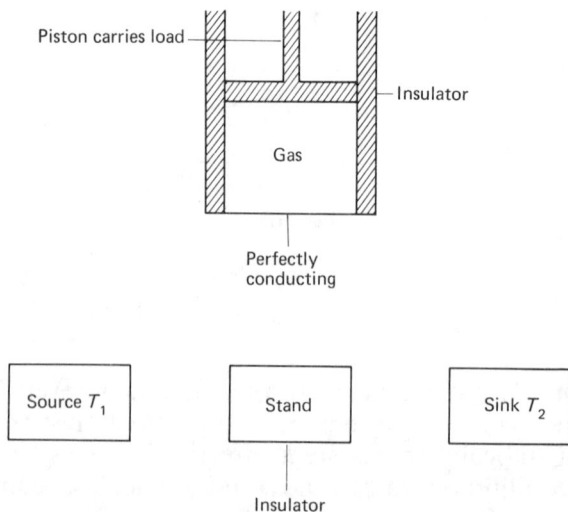

Fig. 11.4

to be a gas, the processes involved in a Carnot cycle may be mechanical, chemical, electrical or magnetic.

The Carnot refrigerator

Since a Carnot cycle consists of reversible processes it may be performed in reverse, in which case it is a refrigeration cycle. Fig. 11.5 shows a Carnot engine, and the engine run in reverse as a Carnot refrigerator. The work W, Q_1, and Q_2 for the refrigerator, are numerically equal to the same quantities for the engine.

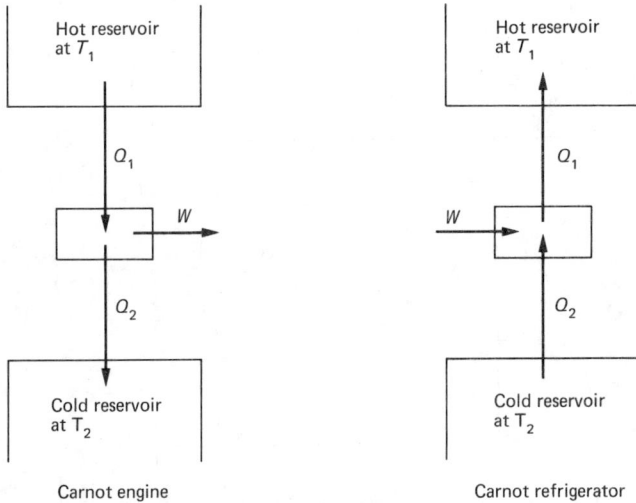

Fig. 11.5

CARNOT'S THEOREM AND COROLLARY

No engine operating between two given reservoirs can be more efficient than a perfectly reversible engine (i.e. a Carnot engine) operating between the same two reservoirs.

PROOF

Imagine a Carnot engine R and any other engine I working between the same two reservoirs and adjusted so that they both deliver the same amount of work W per cycle. The efficiency of the Carnot engine is $\eta_R = W/Q_1$, and that of the engine I is $\eta_I = W/Q_1'$, where Q_1 and Q_1' are the heats taken from the hot reservoir by the Carnot engine and the engine I respectively. Suppose that $\eta_1 > \eta_R$, i.e. $Q_1 > Q_1'$. Now let the engine I drive the Carnot engine backwards as a Carnot refrigerator, as shown in Fig. 11.6. Since the work W required to drive the refrigerator is derived from the engine I, we now have a self-acting device. The net heat delivered to the hot reservoir is $Q_1 - Q_1'$, and the net heat taken from the cold reservoir is $Q_1 - W - (Q_1' - W) = Q_1 - Q_1'$, a positive quantity since $Q_1 > Q_1'$. Thus the sole effect of this self-acting device is to transfer $Q_1 - Q_1'$ units of heat from a cold to a hot reservoir. Since this violates the second law (Clausius' statement) our assumption that $\eta_1 > \eta_R$ must be false, and in fact $\eta_1 \leqq \eta_R$.

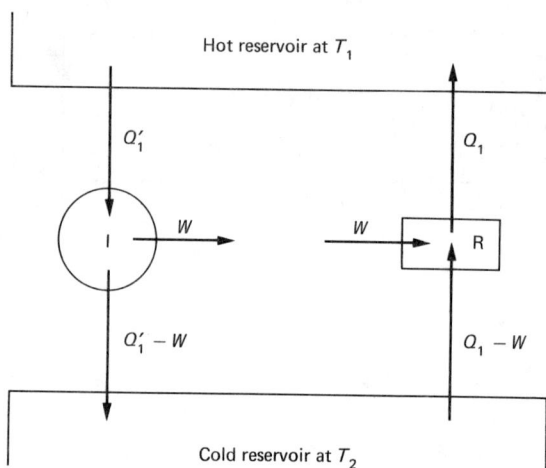

Fig. 11.6

COROLLARY

All Carnot engines working between the same two reservoirs are equally efficient.

If we have two Carnot engines R_1 and R_2 operating between the same two reservoirs and we let R_1 drive R_2 backwards, then Carnot's theorem gives $\eta_{R_1} \leqq \eta_{R_2}$. On the other hand if we let R_2 drive R_1 backwards, then Carnot's theorem gives $\eta_{R_2} \leqq \eta_{R_1}$. Hence $\eta_{R_1} = \eta_{R_2}$. It follows from this result that the efficiency of a Carnot engine does not depend in any way on the working substance.

11.3 The thermodynamic scale of temperature (Kelvin temperature scale)

Since the efficiency of a perfectly reversible engine is independent of the nature of the working substance and depends only on the temperatures of the source and sink, we define a thermodynamic scale of temperature as follows:

$$\frac{T_1}{T_2} = \frac{Q_1}{Q_2}$$

where T_1 is the temperature of the hot reservoir on the Kelvin scale, T_2 is the temperature of the cold reservoir on this scale, and Q_1 and Q_2 are the heats transferred in the Carnot cycle. Now the efficiency of a Carnot engine working between temperatures T_1 and T_2 is

$$\eta = \frac{Q_1 - Q_2}{Q_1} = 1 - \frac{Q_2}{Q_1}$$

$$= 1 - \frac{T_2}{T_1} = \frac{T_1 - T_2}{T_1}$$

If T_2 is zero $\eta = 1$. Thus absolute zero is defined as the temperature of the cold reservoir in an engine with unit efficiency.

To complete the definition of the Kelvin scale, we assign the arbitrary value 273·16 K to the triple point of water, i.e. $T_3 = 273\cdot16$ K and

$$\frac{Q}{Q_3} = \frac{T}{T_3}$$

or

$$T = 273\cdot16\,\frac{Q}{Q_3}\,\text{K}$$

where Q is the heat absorbed from the hot reservoir at temperature T K, and Q_3 is the heat rejected to the cold reservoir at 273·16 K. The corresponding equation for the ideal gas temperature is

$$\theta = 273\cdot16\,\frac{\lim_{P\to0}(PV)}{\lim_{P\to0}(PV)_3}$$

By comparison Q plays the role of a thermometric property.

11.4   The relation between the Kelvin scale and the ideal gas scale

Consider one kilogram mole of ideal gas and go through a Carnot cycle. $PV = R\theta$ defines temperature θ on the ideal gas scale. In Fig. 11.7 we have the following parts making up the cycle: 1–2, isotherm at ideal gas temperature θ_1; 2–3, adiabatic curve; 3–4 lower isotherm at ideal gas temperature θ_2; 4–1, adiabatic curve back to original state.

Considering the isotherms the first law gives

$$dQ = dU + PdV$$

and

$$dU = \left(\frac{\partial U}{\partial T}\right)_V dT + \left(\frac{\partial U}{\partial V}\right)_T dV$$

The first term is zero for an isotherm and the second term is zero for an ideal gas. Thus $\Delta U = 0$ along an isotherm.

The heat input Q is, in this case, equal to the work done, $\int PdV$. Therefore Q_1, the area under the curve between points 1 and 2 is

$$Q_1 = \int_{V_1}^{V_2} PdV = R\theta_1 \int_{V_1}^{V_2} \frac{dV}{V} = R\theta_1 \log_e \left(\frac{V_2}{V_1}\right)$$

Similarly, we find

$$Q_2 = R\theta_2 \log_e \left(\frac{V_3}{V_4}\right)$$

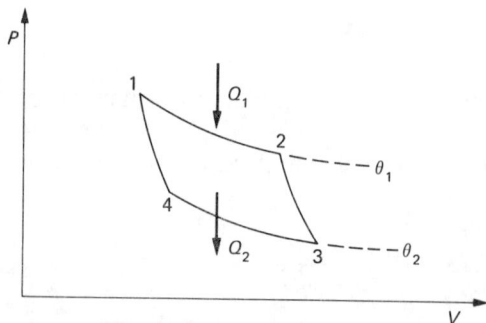

Fig. 11.7

Thus

$$\frac{Q_1}{Q_2} = \frac{\theta_1 \log_e (V_2/V_1)}{\theta_2 \log_e (V_3/V_4)}$$

But, for the isotherms

$$P_1 V_1 = P_2 V_2 \quad P_3 V_3 = P_4 V_4$$

Therefore

$$\frac{V_1}{V_2} = \frac{P_2}{P_1} \quad \text{and} \quad \frac{V_4}{V_3} = \frac{P_3}{P_4}$$

and for the adiabatic curves

$$P_2 V_2^\gamma = P_3 V_3^\gamma$$
$$P_1 V_1^\gamma = P_4 V_4^\gamma$$

Dividing these expressions we have

$$\left(\frac{P_2}{P_1}\right)\left(\frac{V_2}{V_1}\right)^\gamma = \left(\frac{P_3}{P_4}\right)\left(\frac{V_3}{V_4}\right)^\gamma$$

and eliminating the P's gives

$$\left(\frac{V_2}{V_1}\right)^{\gamma-1} = \left(\frac{V_3}{V_4}\right)^{\gamma-1}$$

Therefore

$$\frac{Q_1}{Q_2} = \frac{\theta_1}{\theta_2} \equiv \frac{T_1}{T_2}$$

where T denotes Kelvin temperature.

Thus $\theta \propto T$. To make the scales identical, we make the arbitrary assumption that the triple point of water is $273 \cdot 16$ degrees, i.e. $T_3 = \theta_3 = 273 \cdot 16$. The thermodynamic scale is realised in practice by using the gas scale over the range for which this is possible.

11.5 Clausius' theorem

Fig. 11.8 (a) shows a reversible cycle on a work diagram. Fig. 11.8 (b) shows the cycle divided into strips by a number of adiabatic lines. A reversible zig-zag path consisting of alternate

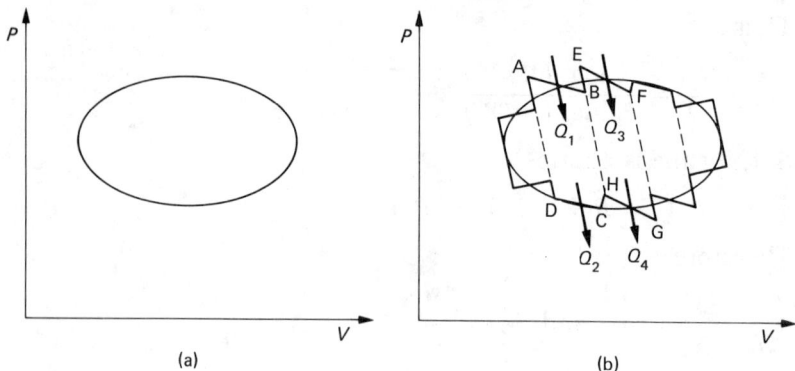

Fig. 11.8

adiabatic and isothermal portions is then drawn, such that the heat transferred in all the isothermal portions equals the heat transferred in the original cycle.

Now ABCD is a Carnot cycle, AB at temperature T_1, and CD at temperature T_2. Thus $Q_1/T_1 = Q_2/T_2$ where Q_1 is the heat absorbed during isothermal process AB, and Q_2 is the heat rejected during isothermal process CD. Taking Q positive for heat absorbed and negative for heat rejected we may write

$$\frac{Q_1}{T_1} + \frac{Q_2}{T_2} = 0$$

where Q_2 is a negative quantity. Similarly considering the Carnot cycle EFGH we may write

$$\frac{Q_3}{T_3} + \frac{Q_4}{T_4} = 0$$

If an equation like this is written for each Carnot cycle and the equations added, we have

$$\frac{Q_1}{T_1} + \frac{Q_2}{T_2} + \frac{Q_3}{T_3} + \frac{Q_4}{T_4} + \cdots = 0$$

i.e.

$$\sum \frac{Q}{T} = 0$$

where the summation is over the complete zig-zag cycle (no heat is transferred during the adiabatic portions). As we increase the number of adiabatic curves, the zig-zag path approximates more and more closely to the original cycle. Hence in the limit we have

$$\oint \frac{\mathrm{d}Q}{T} = 0$$

and this is true for any reversible cycle. This result is known as Clausius' theorem.

11.6 Entropy

With reference to Fig. 11.9, A is an initial equilibrium state and B a final equilibrium state of a thermodynamic system. Suppose that the system is taken from A to B along reversible path R_1, and then from B back to A along reversible path R_2. Then Clausius' theorem gives

$$\oint_{(R_1 R_2)} \frac{\mathrm{d}Q}{T} = 0$$

Therefore

$$\int_{\substack{A \\ (R_1)}}^{B} \frac{\mathrm{d}Q}{T} + \int_{\substack{B \\ (R_2)}}^{A} \frac{\mathrm{d}Q}{T} = 0$$

Hence

$$\int_{\substack{A \\ (R_1)}}^{B} \frac{\mathrm{d}Q}{T} = -\int_{\substack{B \\ (R_2)}}^{A} \frac{\mathrm{d}Q}{T} = \int_{\substack{A \\ (R_2)}}^{B} \frac{\mathrm{d}Q}{T}$$

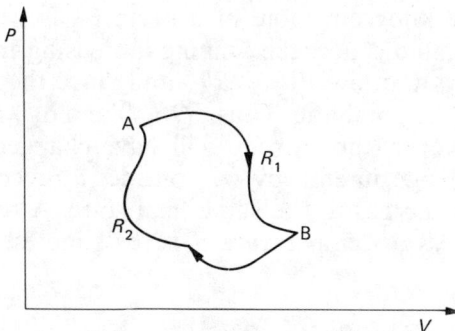

Fig. 11.9

Since the reversible paths R_1 and R_2 are arbitrary, it follows that

$$\int_A^B \frac{dQ}{T}$$
(R)

is independent of the reversible path connecting A and B. It therefore follows that there exists a function S say, such that

$$\int_A^B \frac{dQ}{T} = S_B - S_A$$

This is of great significance. In changing the state of a system from $P_1 V_1 T_1$ to $P_2 V_2 T_2$ we are interested in the heat exchanged Q and the work done W. Unfortunately Q and W cannot be handled simply because both depend on the path taken between the initial and final states. However, we have now found two functions, U and S, which are independent of path and which are amenable to mathematical analysis such as was developed on page 181.

The function S is called the entropy, $S_B - S_A$ being the change in entropy. Also we may write

$$\left(\frac{dQ}{T}\right)_{reversible} = dS$$

and this is the mathematical formulation of the second law of thermodynamics. Note that dQ must be transferred reversibly, and that dS is an exact differential since it is the differential of an actual function. dQ and dW are not exact differentials.

The unresisted expansion of a perfect gas
Consider one kilogram mole of a perfect gas which expands into an evacuated space, the volume increasing from V_1 to V_2. The gas obeys the law $PV = RT$, and since the expansion is sudden it will be adiabatic. Thus $dQ = 0$ and $dQ/T = 0$ for the process. However, the entropy will have changed and to calculate the change in entropy we consider a reversible change which brings the gas to the same final state. A reversible isothermal process at temperature T is the simplest to consider. For this

$$\Delta Q = \Delta W = P dV = RT \frac{dV}{V}$$

and

$$\Delta S = R \frac{\mathrm{d}V}{V}$$

Thus the change in entropy is

$$R\int_{V_1}^{V_2}\frac{\mathrm{d}V}{V} = R \log_e \left(\frac{V_2}{V_1}\right)$$

which is positive since $V_2 > V_1$.

Evaluation of entropy changes
Consider a reversible process for which the first law gives

$$\mathrm{d}Q = \mathrm{d}H - V\mathrm{d}P$$
$$= C_p\mathrm{d}T + \left(\frac{\partial H}{\partial P}\right)_T \mathrm{d}P - V\mathrm{d}P$$

At constant pressure we have $\mathrm{d}Q = C_p\mathrm{d}T$ and $\mathrm{d}S = C_p\,\mathrm{d}T/T$.
Thus

$$\Delta S = C_p \int_{T_1}^{T_2}\frac{\mathrm{d}T}{T} = C_p \log_e \left(\frac{T_2}{T_1}\right)$$

Now for a perfect gas $(\partial H/\partial P)_T = 0$ and hence

$$\mathrm{d}Q = C_p\mathrm{d}T - V\mathrm{d}P$$

Therefore

$$\mathrm{d}S = C_p \frac{\mathrm{d}T}{T} - \frac{V}{T}\mathrm{d}P$$

and

$$\Delta S = C_p \int_{T_1}^{T_2}\frac{\mathrm{d}T}{T} - R \int_{P_1}^{P_2}\frac{\mathrm{d}P}{P}$$
$$= C_p \log_e \left(\frac{T_2}{T_1}\right) - R \log_e \left(\frac{P_2}{P_1}\right)$$

Therefore, for a perfect gas,

$$S = C_p \log_e T - R \log_e P + \text{constant}$$

Going back to the first law and writing it in terms of $\mathrm{d}U$ we have, for a reversible change,

$$dQ = dU + PdV$$

$$= C_V dT + \left(\frac{\partial U}{\partial V}\right)_T dV + PdV$$

At constant volume this becomes

$$dQ = C_V dT \quad \text{and} \quad dS = C_V dT/T$$

Thus

$$\Delta S = C_V \int_{T_1}^{T_2} \frac{dT}{T} = C_V \log_e \left(\frac{T_2}{T_1}\right)$$

For a perfect gas $(\partial U/\partial V)_T = 0$ and

$$dQ = C_V dT + PdV$$

Therefore

$$dS = C_V \frac{dT}{T} + \frac{P}{T} dV$$

and

$$\Delta S = C_V \int_{T_1}^{T_2} \frac{dT}{T} + R \int_{V_1}^{V_2} \frac{dV}{V}$$

$$= C_V \log_e \left(\frac{T_2}{T_1}\right) + R \log_e \left(\frac{V_2}{V_1}\right)$$

Thus, for a perfect gas,

$$S = C_V \log_e T + R \log_e V + \text{constant}$$

Referring to Fig. 11.10, suppose we require the entropy change $S_1 - S_0$ for a perfect gas. Starting with

$$dS = C_V \frac{dT}{T} + \frac{P}{T} dV$$

we see that

$$S' - S_0 = R \log_e \left(\frac{V'}{V_0}\right) = R \log_e \left(\frac{V_1}{V_0}\right)$$

since T is constant. Also

$$S_1 - S' = C_V \log_e T_1/T'$$
$$= C_V \log_e T_1/T_0$$

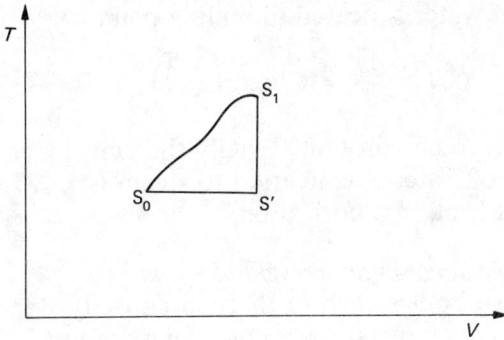

Fig. 11.10

Thus the total change in entropy is

$$S_1 - S_0 = C_V \log_e \left(\frac{T_1}{T_0}\right) + R \log_e \left(\frac{V_1}{V_0}\right)$$

The entropy change from ice to steam
Fig. 11.11 shows the entropy changes which occur as 1 kilogram of ice is heated until it is converted into steam. L_f, the latent heat of fusion, is the heat needed to melt 1 kilogram of ice. The entropy change is thus $L_f/273$. The entropy change when 1

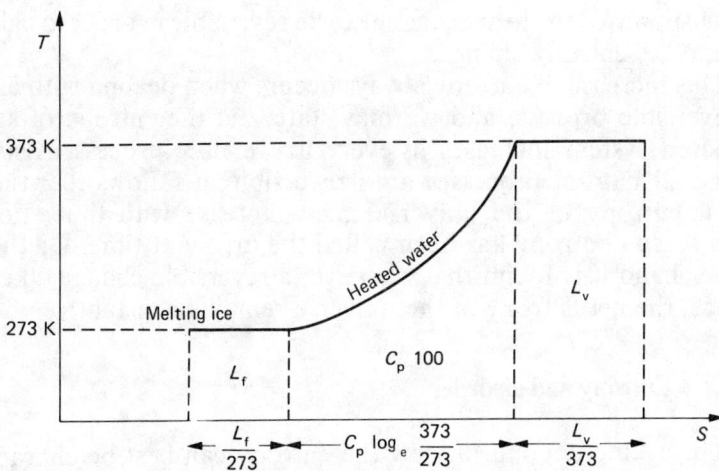

Fig. 11.11

kilogram of water is heated to boiling point is

$$\Delta S = C_p \int_{273}^{373} \frac{\mathrm{d}T}{T} = C_p \log_e \left(\frac{373}{273}\right)$$

assuming C_p to be constant. Finally the entropy change when 1 kilogram of water is converted to steam is $L_v/373$, where L_v is the latent heat of vaporization.

The entropy change in an irreversible process
Consider two bodies each of thermal capacity W, which share heat so that they reach a common temperature T_0, the initial temperature of one being $T_0 - \theta$ and of the other being $T_0 + \theta$. The entropy change of one is

$$W \log_e \left(\frac{T_0}{T_0 - \theta}\right)$$

and of the other is

$$W \log_e \left(\frac{T_0}{T_0 + \theta}\right)$$

a negative quantity. The increase in entropy of the whole system is

$$W \log_e \left(\frac{T_0^2}{T_0^2 - \theta^2}\right)$$

Although the heat exchange in fact takes place irreversibly, we have supposed the heat exchange to be reversible in order to calculate the entropy change.

This increase of entropy always occurs when dealing with an irreversible process, and we may state that the entropy of an isolated system increases in every irreversible process. Also, since all natural processes are irreversible, it follows that the total entropy of the universe must increase with time. For this reason entropy has been called the arrow of time. On the other hand it is found that whenever a reversible change takes place, the net entropy of the universe remains constant.

11.7   Entropy and disorder

From a kinetic point of view the entropy can best be characterized as the measure of the disorder of a system. When we cool a system at constant volume we continuously withdraw heat, and hence entropy, from the system, while at the same time the order increases more and more. As gas becomes liquid and liquid becomes solid the degree of order increases, corresponding to a discontinuous decrease of the entropy.

Consider a metal bar, hot at one end, cold at the other. Some sort of order exists, since most of the high-energy molecules are segregated at one end, while the low-energy molecules are largely at the other end. As a result of spontaneous conduction of heat eventually there is a uniform distribution of energy throughout. Hence a state of partial order has become one of greater disorder; also the entropy of the system has increased. So also, when a gas expands into a vacuum, is there an increase in entropy, corresponding to a state of greater disorder, the most disordered state of the gas being the removal of the walls of the vessel altogether allowing the molecules to spread evenly throughout the whole universe. Moreover, melting the vaporization lead to an increase in disorder and hence an increase in entropy.

As an example of a problem on entropy we will show that latent heats must alway be positive. Consider a phase change, e.g. solid to liquid or liquid to vapour. From what we have said about entropy and disorder, there will be an increase in entropy, the increase being given by L/T, where L is the appropriate latent heat and T is the temperature at which the phase change takes place. Thus L/T must be positive, and since T is positive, then L must always be positive.

Further mathematical development leads to a series of equations called the Maxwell equations. However, the treatment of this subject is not appropriate for this book.

12 Heat Conduction

When dealing with the problem of heat conduction we have to consider:
(1) the transient phase where temperatures are still changing;
(2) the steady state in which temperatures are not changing with time. This state is reached when heat leaves the medium at the same rate at which it is received.

Referring to Fig. 12.1, the definition of thermal conductivity is based on steady-state straight-line heat flow along, say, a rod. The temperature gradient $d\theta/dx$ is uniform, $Q \propto -d\theta/dx$ and also $Q \propto A$. Therefore

$$Q \propto -A \frac{d\theta}{dx} = -kA \frac{d\theta}{dx}$$

where k is the thermal conductivity of the material of the rod.

Consider now the possibility that $\theta = \theta(t)$ and $d\theta/dx = f(x)$, where t denotes time. Referring to Fig. 12.2 in the transient phase, the difference between the heat in and the heat out

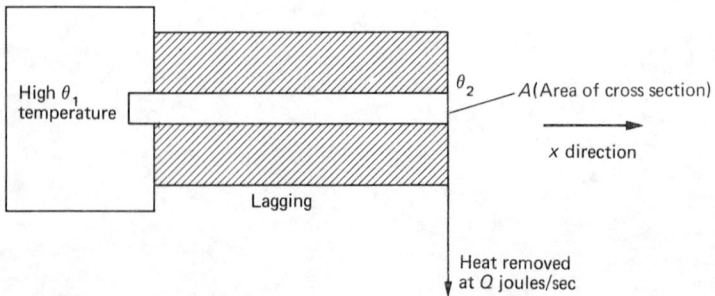

High θ_1 temperature

θ_2 A(Area of cross section)

x direction

Lagging

Heat removed at Q joules/sec

Fig. 12.1

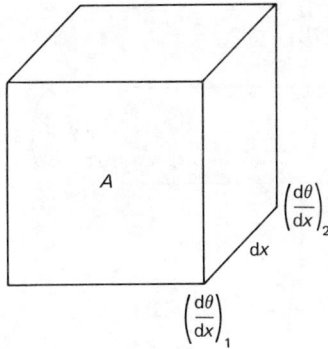

Fig. 12.2

warms up the block; thus

$$d\left(kA\,\frac{d\theta}{dx}\right) = c\,dm\,\frac{d\theta}{dt}$$

where c denotes specific heat, and dm denotes mass. As $dx \to 0$ the temperature of the block becomes more and more precise. But $dm = \rho A dx$, where ρ denotes density. Therefore

$$d\left(kA\,\frac{d\theta}{dx}\right) = \rho cA\,\frac{d\theta}{dt}\,dx$$

and

$$kA\,\frac{d^2\theta}{dx^2}\,dx = \rho cA\,\frac{d\theta}{dt}\,dx$$

or

$$\frac{d\theta}{dt} = \frac{k}{\rho c}\,\frac{d^2\theta}{dx^2}$$

In general

$$\frac{\partial\theta}{\partial t} = \frac{k}{\rho c}\left(\frac{\partial^2\theta}{\partial x^2} + \frac{\partial^2\theta}{\partial y^2} + \frac{\partial^2\theta}{\partial z^2}\right)$$

Consider now straight-line heat flow with lateral leakage (Fig. 12.3). If the rod is thin, for the main part of the rod, we will have straight-line heat flow. Our differential equation now becomes

$$d\left(kA\,\frac{d\theta}{dx}\right) = \rho cA\,\frac{d\theta}{dt}\,dx + \text{lateral leakage}$$

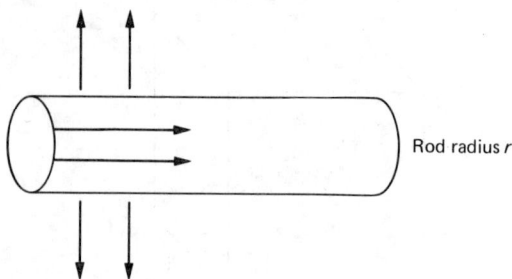

Rod radius *r*

Fig. 12.3

and

$$kA \frac{\mathrm{d}^2\theta}{\mathrm{d}x^2}\mathrm{d}x = \rho cA \frac{\mathrm{d}\theta}{\mathrm{d}t}\mathrm{d}x + \text{leakage}$$

Now assume that the heat lost per second from the walls is

$$= E \times \text{surface area} \times \theta$$

where θ is the excess temperature and E is the emissivity, which depends on the colour and state of polish of the surface, but not on the kind of surface. Thus the heat lost per second from the walls is

$$E.2\pi r \mathrm{d}x.\theta$$

where θ is the temperature excess over room temperature. Thus, we have

$$kA \frac{\mathrm{d}^2\theta}{\mathrm{d}x^2}\mathrm{d}x = \rho cA \frac{\mathrm{d}\theta}{\mathrm{d}t}\mathrm{d}x + E.2\pi r \mathrm{d}x.\theta$$

Thus

$$\frac{\mathrm{d}\theta}{\mathrm{d}t} = \frac{k}{\rho c}\frac{\mathrm{d}^2\theta}{\mathrm{d}x^2} - \frac{E.2\pi r}{\rho cA}\theta$$

For steady-state conditions $\mathrm{d}\theta/\mathrm{d}t = 0$, and so

$$\frac{\mathrm{d}^2\theta}{\mathrm{d}x^2} - \frac{E.2\pi r}{kA}\theta = 0$$

and

$$\frac{\mathrm{d}^2\theta}{\mathrm{d}x^2} - \mu^2\theta = 0$$

This equation has a solution

$$\theta = Ae^{\mu x} + Be^{-\mu x}$$

But $\theta \to 0$ as $x \to \infty$ (the temperature gets nearer and nearer to room temperature the farther one goes along the rod) and at $x = 0$, $\theta = \theta_0$. Thus, using these boundary conditions, we have

$$0 = Ae^{\infty} + 0 \quad \text{i.e. } A = 0$$

and so

$$\theta = Be^{-\mu x} = \theta_0 e^{-\mu x}$$

since $\theta = \theta_0$ at $x = 0$.

Now consider the case of spherical symmetry (Fig. 12.4). In this case we can write

$$d\left(kA\frac{\partial\theta}{\partial r}\right) = \rho Ac dr \frac{\partial\theta}{\partial t}$$

or

$$d\left(k.4\pi r^2 \frac{\partial\theta}{\partial r}\right) = \rho.4\pi r^2 dr.c \frac{\partial\theta}{\partial t}$$

i.e.

$$4\pi k\left(r^2 \frac{\partial^2\theta}{\partial r^2} + 2r \frac{\partial\theta}{\partial r}\right)dr = 4\pi\rho c r^2 \frac{\partial\theta}{\partial t} dr$$

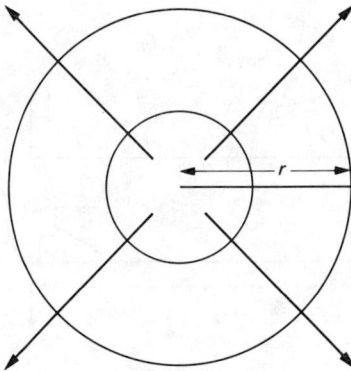

Fig. 12.4

thus

$$k\left(\frac{\partial^2\theta}{\partial r^2} + \frac{2}{r}\frac{\partial\theta}{\partial r}\right) = \rho c\frac{\partial\theta}{\partial t}$$

and so, finally,

$$\frac{\partial\theta}{\partial t} = \frac{k}{\rho c}\left(\frac{\partial^2\theta}{\partial r^2} + \frac{2}{r}\frac{\partial\theta}{\partial r}\right)$$

As a practical application, Lord Kelvin used this equation to obtain an order of magnitude estimate for the age of the earth. Since he was using the equation as it stands and since it is a differential equation, it really only applies to an element. Lord Kelvin applied the equation to surface rocks only. He obtained a rough value for $k/\rho c$ from geological data and values for $\partial^2\theta/\partial r^2$ and $\partial\theta/\partial r$ from measurements of temperatures down mines. For the earth's surface Kelvin found $\partial\theta/\partial t$ was 10^{-7} °C per annum. He was then able to calculate how much time has elapsed since the earth was at 1000 °C say. This time, which gives an order of magnitude for the age of the earth, he found to be approximately 10^{10} years.

Finally, we consider the case of cylindrical symmetry (Fig. 12.5). In this case we have

$$d\left(k.2\pi r.L\frac{\partial\theta}{\partial r}\right) = \rho.2\pi r.L dr c\frac{\partial\theta}{\partial t}$$

thus

$$2\pi k\left(r\frac{\partial^2\theta}{\partial r^2} + \frac{\partial\theta}{\partial r}\right)dr = 2\pi\rho cr\frac{\partial\theta}{\partial t}\,dr$$

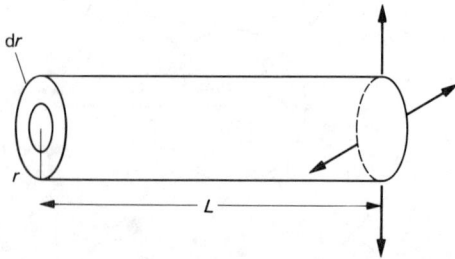

Fig. 12.5

i.e.

$$k\left(\frac{\partial^2\theta}{\partial r^2} + \frac{1}{r}\frac{\partial\theta}{\partial r}\right) = \rho c\frac{\partial\theta}{\partial t}$$

and so

$$\frac{\partial\theta}{\partial t} = \frac{k}{\rho c}\left(\frac{\partial^2\theta}{\partial r^2} + \frac{1}{r}\frac{\partial\theta}{\partial r}\right)$$

12.1 The composite rod for comparing thermal conductivities

Referring to Fig. 12.6 let t, t_2 be the temperatures at the ends of the rod and t_j that at the junction. Let k_1 and k_2 be the thermal conductivities of the two parts of the composite rod, each of length L, Q is the same for all parts of the rod in the steady state. Thus

$$Q = -\frac{k_1 A(t_j - t_1)}{L}$$

$$= -\frac{k_2 A(t_2 - t_j)}{L}$$

Therefore

$$\frac{k_1}{k_2} = \frac{t_j - t_2}{t_1 - t_j}$$

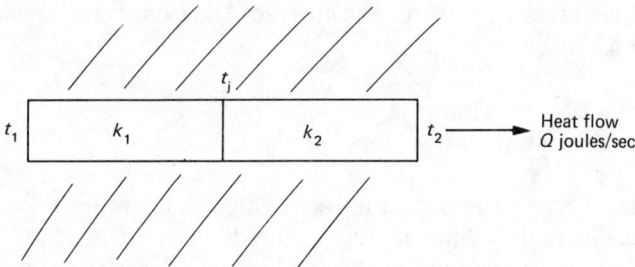

Fig. 12.6

13 Radiation

Thermal radiation is known experimentally to be of the same nature as ordinary light but its wavelength lies in the infrared region. It is emitted by all bodies by virtue of their temperatures and is also given out by closely packed molecules, e.g. liquids or solids.

It is an experimental fact that if one has two bodies, one initially hot the other initially cold, suspended by silk threads in a vacuum, the bodies eventually reach a common temperature. At one time it was thought that cold bodies radiated cold. Another suggestion was that the hot body emits and loses energy and so its temperature decreases, while the cold body receives energy and its temperature increases. This theory had to be rejected since if emission were the only process acting how would the hot body know when the cold body had reached the same temperature so that it should stop emitting? The correct explanation was due to the Swiss Physicist Prévost who gave us his law of exchanges. According to Prévost equilibrium is brought about by two simultaneous processes: both bodies emit and absorb, and at equilibrium they emit and absorb at the same rate.

13.1 Absorptive power

There is some property of a body which determines how much of the thermal radiation that is incident upon it is absorbed. This property is called the absorptive power and depends on the nature of the matter making up the body, the wavelength of the thermal radiation and on the temperature of the body.

If I_λ is the energy per square metre per second incident on a surface per unit waveband, then $I_\lambda \, d\lambda$ is the energy per square metre per second of thermal radiation of wavelength in the range between λ and $\lambda + d\lambda$ incident on the surface. The energy absorbed per square metre per second is $a_\lambda I_\lambda d\lambda$, where a_λ is the absorptive power. For bodies that absorb all the radiation incident on them $a_\lambda = 1$ and for those that do not absorb at all $a_\lambda = 0$. In general black bodies are good absorbers, white bodies are poor absorbers and other bodies range between these two extremes.

13.2 Emissive power

There is some property of a body which determines how much thermal radiation it emits, the hotter it is the more it emits. This property is called the emissive power, e_λ, and depends on the same factors as a_λ. Now, we have that the energy emitted per square metre per second in the wavelength range between λ and $\lambda + d\lambda$ is $e_\lambda d\lambda$.

If a body is in a state of temperature equilibrium the rate of emission of thermal radiation is equal to the rate of absorption. Therefore

$$e_\lambda d\lambda = a_\lambda I_\lambda d\lambda$$

or

$$e_\lambda / a_\lambda = I_\lambda$$

This is known as Kirchhoff's law

13.3 Black-body radiation

Referring to Fig. 13.1, consider an ideal insulated enclosure with several bodies inside. When a state of temperature equilibrium is reached every body is emitting and absorbing thermal radiation at the same rate. Therefore the radiation energy I_λ does not depend on the materials comprising walls or the nature of the matter inside. It seems I_λ is a function of temperature only.

An experiment to check this is the following: consider two enclosures at the same temperature (Fig. 13.2); if I_λ depends on something other than temperature we could have $I_{\lambda_1} > I_{\lambda_2}$. In this case energy could be passed via a suitable filter from one enclosure to the other and we could get work out of a system at

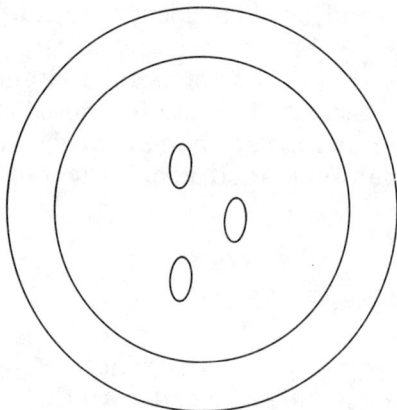

Fig. 13.1

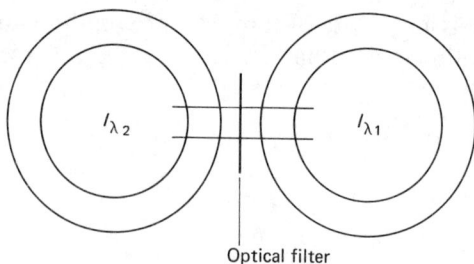

Fig. 13.2

constant temperature. This would contravene the second law of thermodynamics. Hence I_λ is a function of temperature, $f(T)$, only. This radiation is called temperature radiation, cavity radiation, full radiation, or black-body radiation. If we had a body which behaved like a black body for all λ, i.e. if $\alpha_\lambda = 1$ for all λ, then $e_{\lambda_{\text{black}}} = I_\lambda$. In this case one would not need an enclosure; as long as the black body was at a certain temperature one would get the quantity I_λ.

A good experimental approximation to a black body (Fig. 13.3) is an enclosure with a small hole in its wall; incoming radiation is almost totally absorbed. In this case $I_\lambda(=e_\lambda/a_\lambda)$ is a function of temperature only. The total emissive and absorptive powers, e and a respectively, are obtained by integrating

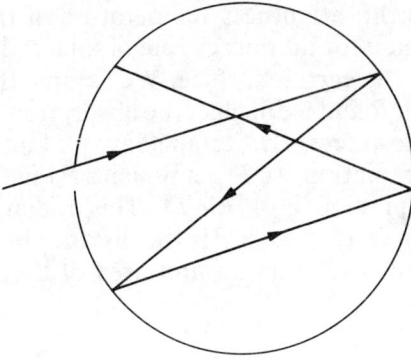

Fig. 13.3

e_λ and a_λ over all wavelengths:

$$\int_0^\infty e_\lambda \, d\lambda = e$$

$$\int_0^\infty a_\lambda \, d\lambda = a$$

The total emissive power of a black body is given by

$$\int_0^\infty I_\lambda \, d\lambda = I$$

Stefan's law

The quantity I has been related to the absolute temperature T by theory and experiment. Stefan measured the radiation from a furnace at a low temperature and at a metal melting point much higher in temperature using a thermopile. It seemed to him that I was proportional to T^4. This relation was later derived theoretically by Boltzmann giving $I = \sigma T^4$ where σ is known as Stefan's constant.

The brightness temperature of the sun

The brightness temperature of the sun is the temperature we obtain when we use Stefan's law to estimate the sun's temperature. In other words we are assuming that the sun emits radiation equivalent to a black body at a particular temperature. If the body is not in thermal equilibrium we can only estimate the temperature.

To calculate the brightness temperature of the sun first we need an expression for the energy rate of solar radiation received on unit area of the earth's surface. We assume that the sun is a sphere of radius R and we neglect the absorption of radiation by the earth's atmosphere. By Stefan's law we know that the rate of emission of radiation is σT^4 per unit area (Fig. 13.4). Thus the total rate of emission is $4\pi R^2 \sigma T^4$. This radiation is emitted equally in all directions so, if d is the distance between sun and earth, we receive radiation on unit area of the earth at a rate given by

$$\frac{4\pi R^2 \sigma T^4}{4\pi d^2} = \frac{R^2}{d^2}\sigma T^4$$

As explained above, the actual rate is not given exactly by this expression since we have ignored absorption by the atmosphere.

The radiation we receive is measured using a thermopile embedded in a thick copper block at room temperature; very thin mica separators are used to avoid short circuiting. For a sensitive thermopile bismuth and silver are used as the pair.

The thermopile is used first to measure the radiation received from the sun (Fig. 13.5); it is then shielded from the sun and heated electrically until the supply of electrical power, W, produces the same reading. When this is so we have

$$W = \frac{R^2}{d^2}\sigma T^4$$

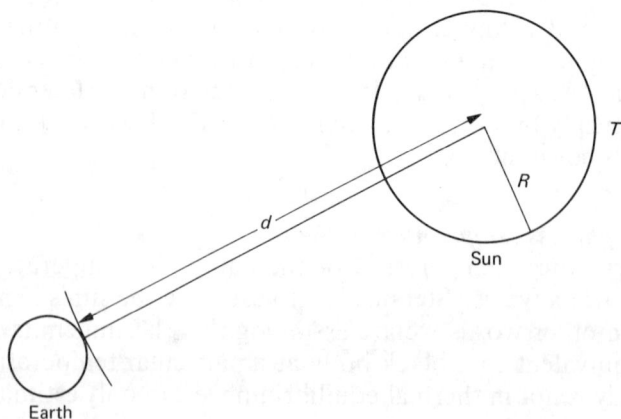

Fig. 13.4

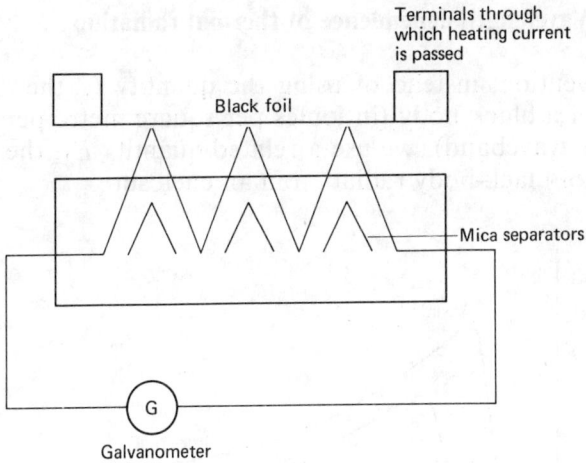

Fig. 13.5

We know that $d/R = 214$, thus we can evaluate T since σ is also known and W is measured.

The brightness temperature of the sun, i.e. the equivalent black-body temperature, obtained by this means is about 6000 K.

Exchange of heat by radiation when surroundings are at a lower temperature

Referring to Fig. 13.6, assume that the body and surroundings are black, and assume also that we have temperature equilibrium. The body is emitting at a rate σT_1^4 and at the same time is receiving at a rate σT_2^4. The body is therefore losing energy at a rate $\sigma(T_1^4 - T_2^4)$. These quantities are measured in units of joules per square metre per second.

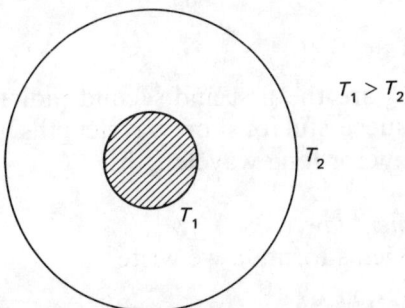

Fig. 13.6

13.4   Wavelength dependence of thermal radiation

By convention, instead of using the quantity I_λ, the emissive power of a black body (in joules per square metre per second per unit waveband), we use a related quantity E_λ, the energy density of black-body radiation in an enclosure.

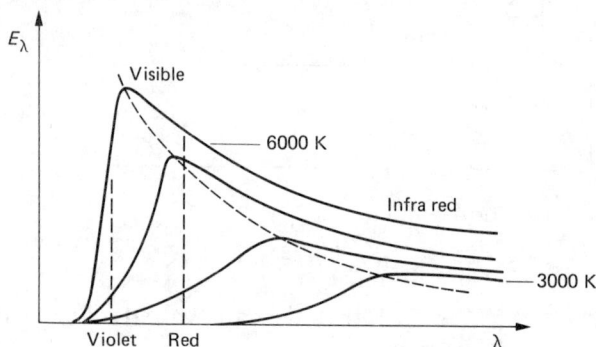

Fig. 13.7

Wien's law, which was obtained from thermodynamical considerations, is (see Fig. 13.7)

$$E_\lambda = \lambda^{-5} f(\lambda T)$$

The same type of curves are obtained if I_λ, emissive power, is plotted against λ. Each curve has a maximum which shifts towards smaller values of λ as the temperature increases. This agrees with the fact that as the temperature of an incandescent body is raised the body becomes brighter and whiter in appearance. Wien's formula for I_λ, the emissive power of a black body, is

$$I_\lambda = C_1 \lambda^{-5} e^{-C_2/\lambda T}$$

where C_1 and C_2 are the first and second radiation constants. This formula is successful for short wavelengths but gives values which are too low for long wavelengths.

Wien's displacement law
Starting from Wien's formula we write

$$I_\lambda = C_1 \lambda^{-5} e^{-C_2/\lambda T}$$

Differentiating this expression gives

$$\frac{\mathrm{d}I_\lambda}{\mathrm{d}\lambda} = -5C_1\lambda^{-6}e^{-C_2/\lambda T} + \frac{C_1 C_2}{\lambda^2 T}\lambda^{-5}e^{-C_2/\lambda T}$$

For a maximum $\mathrm{d}I_\lambda/\mathrm{d}\lambda = 0$ and so

$$5C_1\lambda_{\max}^{-6}e^{-C_2/\lambda_{\max}T} = \frac{C_1 C_2}{\lambda_{\max}^2 T}\lambda_{\max}^{-5}e^{-C_2/\lambda_{\max}T}$$

Therefore

$$\lambda_{\max}T = C_2/5 = \text{constant}$$

Hence

$$\lambda_{\max}T = \text{constant}$$

is Wien's displacement law. At ordinary temperatures the maximum occurs in the infrared region. Radiation from the sun has a maximum at 4500 Å which is in the green part of the visible spectrum. Our eyes have maximum sensitivity to this wavelength. If we know what the relation $E_\lambda = \lambda^{-5}f(\lambda T)$ is at one temperature then we can deduce the shape of the curve at any other temperature. To see this consider Fig. 13.8: first we choose λ_1 and E_1, then we choose λ_2 such that $\lambda_2 T_2 = \lambda_1 T_1$. Now, we have

$$E_\lambda = \lambda^{-5}f(\lambda T) = T^5/T^5\lambda^{-5}f(\lambda T)$$
$$= T^5 F(\lambda T)$$

where F is another function. F is the same for both curves since

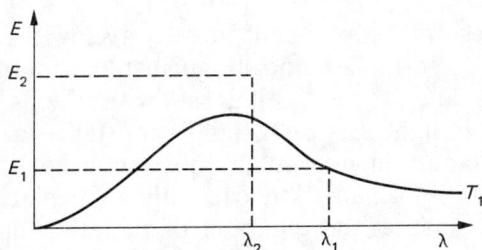

Fig. 13.8

$\lambda_1 T_1 = \lambda_2 T_2$ therefore we have

$$\frac{E_2}{E_1} = \frac{T_2^5}{T_1^5}$$

and so we can deduce E_2 since we know E_1, T_1 and T_2. This procedure can be performed successively for all choices of λ.
We have seen that Wien's formula is

$$I_\lambda = C_1 \lambda^{-5} e^{-C_2/\lambda T}$$

About the same time (1900) Rayleigh and Jeans applied the principle of equipartition of energy to a system of electromagnetic vibrations of different frequencies and produced a formula

$$E_\lambda = 8\pi\lambda^{-4} kT$$

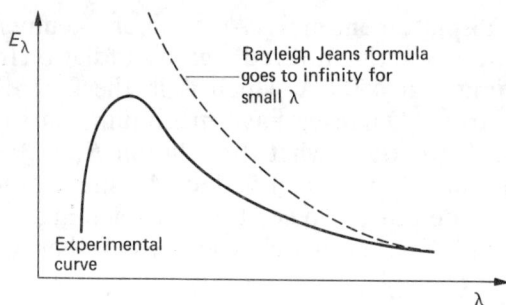

Fig. 13.9

The Rayleigh–Jeans curve (Fig. 13.9) agrees well with experiment for long wavelengths (in the infrared) but predicts no maximum and is completely wrong in the ultraviolet.
Planck finally succeeded in finding a formula

$$E_\lambda = \frac{8\pi hc\lambda^{-5}}{e^{hc/k\lambda T} - 1}$$

which agrees with experiment at all wavelengths. When λ is large Planck's formula is indistinguishable from the Rayleigh–Jeans result, and for small λ it agrees with Wien's formula. c is the velocity of light, k is Boltzmann's constant and h is Planck's constant. Planck introduced the novel concept that emission and absorption of radiation can only take place in definite amounts or quanta, the quantum of radiation having energy $\epsilon = h\nu$ where ν denotes frequency, i.e. $\epsilon = hc/\lambda$. This was the starting point of the quantum theory.

Index